FALLING BEHIND

HOW CHINA OVERTOOK USA IN THE FIGHT AGAINST CLIMATE CHANGE AND HOW USA CAN CATCH UP

FAHD NAZIR

ACKNOWLEDGMENT

The completion of this undertaking could not have been possible without the encouragement and support of so many people whose names may not all be enumerated. Their reassurance is sincerely appreciated and gratefully acknowledged. However, I would like to express my deep appreciation and indebtedness to my parents:

Nazir Ahmed Khan and Rehana Tehseen for their endless support and encouragement during this endeavor, without whom this book would not be possible. To all relatives and friends who in one way or another extended their support either morally or otherwise. Thank you!

Above all, thanks to the Great Almighty, the author of knowledge and wisdom, for his countless blessings and love.

Finally, I would like to dedicate this book to all who are genuinely concerned about the environment and climate change. I encourage you to join my Facebook group below and share your ideas and feedback. I would be interested in hearing your feedback and constructive criticism.

www.facebook.com/groups/climatechangefahd/

Thank you,

Fahd Nazir

CONTENTS

Introduction 7

1. China's Environmental Woes 23
2. The Turning Points 51
3. China's Environmental Goals 69
4. Collection and Access to Information 79
5. Provision of and Access To Funds 91
6. Tighter Regulations 111
7. China's Environmental Status Today 129
8. Overtaking the US 145
9. Learning From China 161

Final Words 181
References 187

INTRODUCTION

"Tackling climate change is a shared mission for mankind ... Let us join hands to contribute to the establishment of an equitable and effective global mechanism on climate change, work for global sustainable development at a high level and bring about new international relations featuring win-win cooperation."

— XI JINPING, CHINESE PRESIDENT

There is a war brewing on the surface of the international political arena, a veritable kettle of hot water surrounding issues of a potential trade or Cold

War between the US and China. The reasons are manifold and date back years—some being economic competition, concerns relating to international security, and problems surrounding territory and demography.

However, the looming issue is not built around these problems but is rather centered on one crucial question: who will emerge victorious in the marathon sprint to achieve a net-zero carbon emission footprint?

For decades now, we have known that a climate war has been simmering under a vat of international political discontent. Competition and disputes surrounding the climate are nothing new, but climate change increasingly seems to be veering on the edge of emerging as the singularly dominant economic and political face of international relations. What's *more* interesting? In the race between the US and China to emerge as champions of the environment, the US is falling far behind.

The fragile relationship between China and the US is the stuff of political lore, and this fragility is based on competition over a number of arenas, including technological progress, capital investment, and cross-border trade. If this wasn't enough, climate change is also becoming a significant ground for competition. Take, for example, this statistic: China spent almost double of what the US did on energy-transition technologies in the last decade.

The domination of the distribution network, domestic industrial standards, human rights legislation, and carbon levies on international trade are all examples of imperative competitive zones. The situation is fragile enough to warrant a statement from the Bank of America's managing director, according to whom (and I quote), a "*climate war*" is brewing in the horizon between the two capital cities (Washington and Beijing), and this will become more and more apparent as climate change is poised to become the most important financial and political issue over the next few twenty years (Smith, 2021).

The CNBC's report on the political and economic battle based on environmental issues also states that China has taken the pledge to reach *net-zero* carbon emissions by 2060. As per a recent equities research study by Goldman Sachs, the countries which currently have net-null commitments account for little below 50% of all global emissions of greenhouse gases like carbon dioxide, China comprising around two-thirds of these.

During the Clinton administration, the US trade representative, Harry Broadman, has even been quoted as stating that major economies around the world will need to evaluate and improve their research and development strategies and increase scientific and technological collaborations so as to stand a competitive edge against China.

This report also brings something very pertinent to light: strategies to preserve the climate aren't just based on altruistic environmental improvement conditions like "*if we do this, our future generations will benefit.*" For sure, that is a nice thought, but in terms of real-life politics, there are much bigger players at stake because the economic implications of climate improvement and energy transition could surpass a sixty-nine-trillion-dollar investment in this century and a four trillion annual dollar investment per year, respectively (Smith, 2021).

That is so much cash flow that even someone as seasoned as me is overwhelmed.

Why is the environment suddenly a watchword for everyone? We often do not give enough credit to how fragile the cords of existence are. Take humanity, for instance. The very core of our existence is mandated on the holistic survival of all living species on the earth. Everything that we see around us is tied by a delicate balance—on days of unforgiving heat, some crops come to fruition. If this unrelenting summer is replaced by an untimely heavy monsoon, these crops are ruined. This affects produce, which in turn affects the yields of farms, which in turn hampers national produce and disturbs the delicate balance that is tying us to Mother Earth.

All of us exist as part of a global Biosphere, which is a term that we owe to Vladimir Vernadsky. Vernadsky was a Russian scientist who was in his prime during the 1920s. He coined the term biosphere to refer to a symbiotic whole—a complete ecological environment encompassing the globe, in which each and every creature is dependent on one another. We have individual parts within this bigger whole, like ocean ecosystems, forest ecosystems, desert ecosystems, and more.

Now, think about our day-to-day existence. We rely on a number of things, which include living and non-living items. We need water to drink, money to buy items of sustenance, food to eat, shelter, and clothes. These are all non-living things, many of which owe their origins to living things (for example, wool, which is used to make sweaters, comes from sheep, and the food we eat is derived from plants and animals). Broadly, the non-living items that sustain us include water, air, soil, and nutrients. The living items include plants, animals, microbes, and the company of other humans.

The foundations that sustain all species inside a tiny living ecosystem will be present when the ecosystem is thriving since it will include all nutrients and chemical components in a single cycle. When these species produce food and consume it, they cycle elements, and even when they die, they cycle elements again, allowing for new components to take their place and bring about constructive change. This cycle provides us with access to everything we need to stay alive.

We're reliant on ecosystems for numerous functions, including climate regulation, which involves, among other things, water, air, and soil upkeep. Ecosystems provide food and also take care of nutrient recycling. Raw ingredients and other necessary resources are provided for manufacturing pharmaceuticals and other comparable products. They are an integral part of our global commerce, and the world could not function without them. Environmental degradation and pollution can lead to various ailments, including respiratory difficulties and cancer. One could contract typhoid, dysentery, and more by being exposed to water contamination. Lack of access to food would lead to famines and starvation.

There is no dearth to the difficulties our species and all other species will face if our environment is not protected. Of course, every small, individual step matters. But what is of utmost importance is that political authorities should advance environmental stewardship.

Julia Marton-Lefevre, the director-general of the IUCN (*International Union for the Conservation of Nature),* the environmental advisory body to the United Nations, maintains that other issues often sideline the political leaders of the world, and because of this, they sometimes forget how essential it is to conserve our environment in order to ensure the wellbeing of global society (Millman, 2014). This doesn't just mean preserving bits and bobs of the environment because they are "pretty to look at" or are potential/existing "tourist attractions," but rather, preserving the complete whole.

The USA is among the ultra-developed nations of the World. It is a dream for millions to make a living and a home there. With this being said, it has a disappointing track record when it comes to preserving the environment.

It has not only repeatedly reversed course on its internal environmental protection policies, but it has also gravely harmed the integrity, effectiveness, and performance of global politics of environment protection as a result of these actions. Until it can adequately explain its actions in the environmental sphere to its own citizens and the global population, the US will not be justified in the eyes of its own inhabitants and of the world at large.

Consider this, for example. Globally, scientists by a landslide majority have maintained that global warming, which is a severe issue harming our environment, is majorly the result of human activities. The world recorded its warmest decade between 2011-2020, and the global average temperature rose by 1.1 degrees Celsius as compared to pre-2019.

Every decade, reports estimate that global warming induced by human beings increases by 0.2 Celsius. Rises in temperature have negative impacts on both human health and safety and the overall natural environment. This means elevated dangers of environmental changes, which include the excess release of heat entrapment-prone gases like carbon dioxide, which in turn leads to more global warming.

Since the mid-1700s, our world has released above one and a half trillion tons of carbon dioxide. Can you imagine how much damage that has caused to our environment? A logical argument that follows is that the mega-superpowers, nations like the US, which are responsible for a majority of these emissions, should be given the maximal responsibility of working to reduce them.

Statistics, however, point to a whole other story. Data gathered by Hannah Ritchie for *Our World in Data* in 2019 shows that the USA has been responsible for more carbon dioxide emissions than any other country in the world till now—clocking in at a whopping four hundred billion tons since the early 1750s. Historically, the global weight of 25% of all emissions is on its huge shoulders. *This is twice* more than what has been done by China (Ritchie, 2019).

In 2018, the US National Climate Assessment, comprising three hundred scientists of repute, plus agents of the federal government, reported that human activities, particularly the emission of greenhouse gases like carbon dioxide, methane, and nitrous oxide (which are incidentally produced by human activities like industrial production, generating fuel, and burning waste) are the main factor behind global warming.

Again, another report published in 2014 by the *IPCC (Intergovernmental Panel on Climate Change)* clarified that human activities are solely responsible for releasing excessive greenhouse gases that pollute the environment.

Now, what does the policy have to say on this? Political activists in the USA have created an atmosphere of significant uncertainty on the current debate on climate change, and they would rather view global warming not as an issue caused by human beings but rather *as a natural phenomenon over which we have no control.* In the 112th US Congress, 74% of the Republicans in the Senate openly stated their disbelief of the scientific evidence (which is monumental) behind global warming.

To the extent that a researcher went on to note, and I quote here, *"With each passing year it appears that there are fewer Republican members of Congress willing to embrace or act upon scientific knowledge, particularly when it relates to issues such as climate change"* (Gropp, 2011).

Trump's government has continually dubbed global warming a conspiracy, and as a result, his allies refuse to accept that climate change is a reality. The Trump administration not only moved to overturn the Former presidential administration's Clean Power Plan, but also has consistently sought to eliminate environmental measures that were introduced by President Obama and his administration. Trump's administration has persisted in making the fossil fuel industry more competitive by revoking executive branch policy regarding climate change.

Over seventy significant environmental regulations have been changed, repealed, or substantially revised since Trump came to power. This potentially translates to a substantial increase in greenhouse gas emissions, as well as an increase in the deaths caused by air pollution.

On the other end of the world, China continues to make significant progress in terms of improving the quality of the environment.

We will go into details about this in the forthcoming chapters, but for now, this should suffice: the Environmental Protection Law of the PRC (*People's Republic of China)* has laid forth the structure for ensuring that between economic developments and social upliftment, the environment does not get left behind. This law is the cardinal law responsible for environmental protection in China. It has laid down clear definitions of the rights and duties of both the government and the people of China concerning environmental protection.

China has also worked to establish a number of laws to protect the environment and improve the natural resources necessary for environmental protection. These include the Laws for preventing and controlling water and air pollution, pollution by solid wastes, laws to contain marine environmental degradation, laws to preserve forests, grasslands, fisheries, and more.

Additionally, it has brought into effect the Environment and Resources Protection Committee, responsible for creating and evaluating laws protecting the environment and natural resources. The national government has also supported the creation of province and city-based environment and resource protecting organizations.

When you think about it, *how much can one person do?* Global environmental improvement is a collective effort, and the ones in power have to be held accountable to bring structural alterations. This means that we need to understand what the USA is doing wrong, what it should be doing more of, and what it should stop immediately.

This brings me to why this book is vital for you, dear reader. Today, if you have come across this book, you are

- An environmentalist who is concerned about climate change and the future of the planet,
- Interested in learning more about what can be done on a broader scale to address climate change, such as recycling, buying secondhand, composting, or going zero waste,
- An American or based in the United States,
- A professional, or in possession of a white-collar occupation, or/and
- Part of a young family or are considering beginning a family in the near future.

You have witnessed the constant degradation of the environment while residing in the States, and despite the apparent fact that the importance of human efforts and lifestyle adjustments is mandatory in combating climate change, you are also sensitive to the notion that this will never be sufficient until systemic changes are implemented in the form of laws, regulations, and incentives/sanctions.

You are rightfully dissatisfied with the lack of action on the side of the United States to combat climate change, and you expect more accountability from the governmental authorities. To know more, you understand that you have to learn about what other countries are doing that could benefit the United States as well.

This is where I come in. I am a graduate of the illustrious Thunderbird School of Global Management, with a Masters in International Business. I have a history of being part of professional circles which are generally not known for qualities like environment management—I will admit that I have worked in the Oil and Gas industry. (*Shocking*, right?)

But, being part of such circles at close quarters makes one more aware of the changes happening in our environment and the effects of our actions on our future generations. I have seen firsthand how destructive businesses can be, and I have insight into what corporations prioritize.

So, I decided to take a step back and spent five years studying and researching climate change, consulting with various experts to enhance my knowledge base.

Today, I am sharing my life's work with you in an effort to help recognize how you can influence government and corporate policies on climate change. This matters deeply to me because it is only as individuals, corporations, and the government working together that we can combat climate change.

I am writing this book because I believe many of us would like to help the environment but have no idea what can be done on a large scale. This book will provide some insight and best practices from China's environmental improvements.

- Through this book, you will gain perspective on the background of China's environmental actions and policies and how they have learned from their past omissions.
- You will read about the turning points that made them adopt new practices and policies.
- From there, we will read about ideas that have been successfully implemented by the Chinese government and can be adapted by the US.

- We will also look at statistics showing China's progress in meeting its climate change goals and study a comparison of China and the US's environmental actions.

Without further ado, let us begin.

1

CHINA'S ENVIRONMENTAL WOES

Here's an interesting bit of information: did you know that China's origins are predominantly rooted in agriculture? For about four thousand years, China was a nation with the spotlight on farmers. When the People's Republic of China came to be established in 1949, almost all land that could be plowed and sowed had been subject to cultivation and the workings of irrigation and drainage systems. China practiced intensive farming because it needed high crop yields to feed its growing populations. China's policymakers have been trying a number of big social concepts since 1949 to increase agricultural output. First and foremost, a major land reform initiative removed landlords and transferred ownership of land to farmers.

Following that, farm families were gradually grouped into cooperatives, collectives, and, ultimately, people's communes to better serve their needs. After more than two decades of experience with communes, authorities decided to dismantle them because they had grown too bureaucratic and inflexible to adapt to the needs of agricultural production, which demanded more flexibility. In addition, agricultural production incentives remained dormant inside the communal structure. China's authorities initiated a campaign of extensive agricultural reforms in 1978, which has continued to this day. Integrated into contractual ties connecting producers to socioeconomic associations and companies were new incentives that were incorporated into the administration of townships and villages.

Since the founding of the People's Republic of China in 1949, the industrial sector has received the vast majority of Chinese investments and administrative resources. Generally speaking, the agricultural sector got particular attention only when the Chinese leadership believed that the industry was starting to impede the country's overall economic growth. In general, agricultural production grew at a rate consistent with population increase, but not at a rate sufficient to improve living standards.

Now, you may be familiar with Mao's Great Leap Forward, which began in 1958. For the sake of lending credit to this chapter, I will explain it briefly. Under the People's Republic of China, economic development ensued with the onset of a hundred and fifty developmental projects, all staffed and financed by Soviet Russia. However, owing to differences in the political ideologies of Chinese leader Mao Zedong and his Russian counterpart Khruschev, there was a break in cooperation between the two nations. This break led to the army's withdrawal of about fifteen thousand engineers and staff that the Russian government had lent to China, and all project blueprints were annulled. At this time, China did not possess the finances or the technological prowess to tend to these projects independently. Zedong became increasingly conscious of how reliant China was on foreign aid. This worry led him to the conviction that China had to industrialize by its own means.

In 1921 Mao was fully aware of the first attempt made by the Soviet Union to build a socialist economy when farmers reacted to the harvest confiscation by refusing to plant and generate as much produce as in the preceding year. Although it may be that Mao was not conscious of it, the collectivization effort in the Soviet Union was a complete failure in production, and as a consequence of this failure, there was a catastrophic famine in Ukraine. Nevertheless, Mao supported integrating Chinese farmers into the municipalities. In other words, Mao essentially reversed the land allocations given to the peasants in the years just following 1949.

First, the farmers were arranged into cooperatives of 20 to 40 families, each representing a single farm. That was at the local level at that time. After, county-wide collectives consisting of hundreds of thousands of individuals took over the previously existing cooperatives. In addition to calling for the creation of the municipalities, Mao urged farmers to build backyard furnaces for iron and steel manufacturing for tools. Peasants had to melt precious items to produce worthless metal mass, which was really the planned result of the operation. This was because the state encouraged the peasants to increase their production from the furnaces of their backyard, and as they ran out of scrap, they started to melt anything they could find to create sculpted metal, including tools and utensils. A small amount of this destruction of important objects might be attributed to juvenile enthusiasm to increase production from backyard blast furnaces, but the bulk is most likely driven by the desire to fulfill production requirements from stoves.

In the Great Leap Forward, the most destructive consequence of home blast furnaces and other companies which did not support China's agricultural reliance was to distract workers from food production, which led to a food crisis.

Moreover, nobody with the ability to change things was informed of the drop in food production due to centralized control, which made the situation worse. Communal leaders felt pressure to produce more than before, and they were obliged to report when production decreased. Instead of telling what the top-level management needed to hear, it reported what it wanted to hear. In this manner, policy errors that led to food shortages persisted long after anything could be done. The central government worsened the problem for farmers by taking a part of the earnings based on bogus production figures and leaving the farmers with inadequate cash.

The Great Leap Forward has many nuances, and if I went into all of them, we would veer off the course we need to discuss. So, let me leave the key points here for you.

The Great Leap Forward resulted from Mao's urgent need to compete with bigger superpowers at the time and showed that China did not need to depend on anyone else for its own internal infrastructure and technological developments.

Part of the movement focused on constructing large-scale infrastructure projects like dams, reservoirs, and irrigation channels. The intention of this was to support backyard furnaces and automated irrigation, and China implemented large-scale factory construction to increase agricultural outputs.

Precisely six hundred thousand backyard furnaces were installed within a year, with the intent to manufacture iron and steel. By the end of 1959, the number of factories in China had more than doubled, from one hundred and sixty thousand to three hundred and ten thousand, a nearly twofold rise.

Which brings us to the environmental consequences of the Great Leap Forward.

As of 1961, there were four thousand power stations, nine thousand cement factories, and eighty thousand agricultural repair shops in operation, all of which were spewing sulfur dioxide directly into the environment, fueled by coal extracted in fifty-nine thousand mines. Additionally, the backyard furnaces can be credited with an excessive release of greenhouse gases like carbon dioxide, in addition to toxic chemical fumes from mining ores (Mac, 2015).

During 1978, China began turning to a more market-oriented socioeconomic structure. The focus of modernization was not limited only to industry or technology alone but extended to agriculture, science, and defense. Chinese Premier Zhou Enlai initiated this drive during the winter of 1964; however, the progress of this drive halted in the face of the Cultural Revolution (1966-1976). As 1978 approached, the grounds for economic reform continued to grow. This was for several reasons; the Cultural Revolution was not popular among the population, and the Chinese Communist Party and the Government had to change course to garner support. Zedong passed away in 1976, and Deng Xiaoping became the leader of the PRC.

Experimentation among economic planners showed many of them that managing and implementing a planning system was problematic and inefficient. The writings of economic authorities that emerged in the late 1970s bear testimony to this because they all felt that the planning system was inherently flawed, albeit to varying degrees of comprehension and different notions of what was the exact cause of the inefficiency.

The massive economic development and growth that occurred in some of the more market-oriented bordering economies offered as models of how a market economy might operate better—examples included nations like Hong Kong, Taiwan, South Korea, and Singapore. Moreover, the disparities in economic success between North and South Korea, East and West Germany, and Eastern and Western Europe served to underline this argument.

As a result of the economic implications of central planning, the Chinese people had also experienced many difficulties, including shortages of consumer products, a lack of variety, and a lack of development in product quality. To get what they needed, they had to bend over backward and kowtow to the manufacturers and distributors of the items. They had to rely on personal relationship dynamics to obtain what was regarded as necessities for daily living.

These causes provided the fertilizer to the already growing grounds for economic reform. Following the adoption of a new socioeconomic road plan by the eleventh Central Committee of the Chinese Communist Party at its Third Plenary Session in late 1978, China's reform and open-door policy began in earnest, with the establishment of the People's Republic of China. The Chinese government began pursuing an active approach in 1978, under the leadership of Deng Xiaoping, who had decided to return to the world of politics after three previous failures. The Chinese government took the position to achieve economic growth by actively inducting foreign capital and technology while maintaining its commitment to collectivism.

Clearly, this policy shift aimed to rebuild the country's culture and economy, devastated by the Cultural Revolution. There was no question about that goal. The policy shift appears to have been motivated by the realization that Chinese income levels were so low in comparison to those in other Asian countries that the continued existence of the Chinese state and the communist government was in jeopardy. Hence, it required a policy shift to raise the Chinese people's living standard through economic development.

Several regions were opened up to foreign investment, including economic zones, large coastal cities, financial and technical innovation zones, delta open zone development, peninsula open zone development (including the opening of border cities), and high-technology industrial development zones. Due to the establishment of these zones, significant sums of foreign investment, mainly from Hong Kong and Taiwan-based companies, started to pour into the region. Chinese officials also promoted the socialist market economy concept, for which they had been advocating for decades. Due to the reforms, there was an explosion of entrepreneurial activity, culminating in the establishment of significant populations of innovators and venture capital companies across China.

Foreign investment, science, and managerial know-how enabled China to harness its vast labor resources and geographical expanse into a vehicle for rapid economic growth. Following the adoption of an accessible strategy in the early 1980s, economic development increased significantly throughout the first half of the decade. The Chinese economy had a period of stagnation in 1989, but by 1990, the economy had rebounded and was witnessing high growth rates once again. The fast growth of the economy was accompanied by a rise in the gross domestic product (GDP) per person (per capita) (GDP). In 1998, per capita income increased by a factor of fourteen times above what it had been in 1980.

During the readjustment phase, the Chinese government tested the main concepts of the structural reforms in a handful of locations to ensure they were functional. What turned out to be the most effective reform effort was the government's recommendation in 1979 that disadvantaged rural divisions in mountainous or arid regions adopt a contractual responsibility system of agricultural output to increase their revenues in order to improve their living conditions. According to the responsibility system, farm-level families were given a chance to generate income from a piece of land in return for providing a set quantity of product to the collective at a defined price for a specified period. As a consequence of this arrangement, farmers were given tremendous incentives to reduce manufacturing costs while simultaneously boosting output. Following its introduction, the accountability system was rapidly adopted by many farm units in a diverse range of socioeconomic areas.

Agriculture had also benefited from the government's guidance and support to implement free farmers' marketplaces in both cities and suburbs, and the capacity for specific households to function as skilled households, concentrating their endeavors on the output of a rare and valuable commodity or service for a profit.

The government actively promoted the establishment of cooperatively owned and managed industrial and service companies to relieve some of the hardship among young people while promoting the development of the supply of light industrial goods. Despite the restrictions placed on individual business during the Cultural Revolution, it was once again possible to find independent enterprise in the city's market arena after having been virtually eradicated during the period of the Cultural Revolution. In preparation for China to be ready to contend on an equal basis with the rest of the civilized world in international commerce, a wide variety of collaborative, commercial, and financial agreements with foreign companies were allowed.

Given the interconnectedness of the industry's production operations and the difficulty of adopting a single-core plan, the accountability system failed to generate the same dramatic improvements as it had in agriculture. A policy-oriented on more openness and wider market involvement, significantly improved opportunities for most companies, resulting in rapid development and improved efficiency. By 1985, the practice of remitting taxes on earnings and keeping the balance had become widespread, boosting the incentive for businesses to maximize profits while also significantly expanding their autonomy.

During Deng Xiaoping's leadership, the Chinese government often said that its fundamental strategy was reform and openness, which meant modernizing its economic structure while simultaneously opening up to international commerce.

The importance of international trade increased significantly as a result of the economic reforms, far surpassing its significance at any other time in history. A decade before the reform era, the total value of imports and exports had seldom exceeded 10% of national GDP. 1980 saw it account for 15% of the population, 1984 saw it represent 21% of the population, 1986 saw it reach 35% of the population. Under Deng Xiaoping, foreign trade was seen as a significant source of investment money and cutting-edge technology, in contrast to previous eras when China was committed to the goal of self-sufficiency.

The government also found that the reform measures, which were expressly designed to invigorate the economy by transferring the primary responsibility of distributing services and goods from the government scheme to the market, had a beneficial effect on the domestic trade sector. Even though private enterprise and free-market operations were allowed and promoted throughout the 1980s, policymakers were constantly at war with local governments' efforts to levy excessive taxes on individual merchants during this period. China's state-controlled network of commercial agencies and retail stores coexisted with a rapidly expanding privatized and collectively owned system that competed with it vehemently, providing Chinese people with a broader range of consumption choices than at any other time in history.

When it comes to results, the period of readjustment had already produced promising ones:

- incomes had increased significantly
- access to food, housing, and other consumer goods had improved
- there was strong growth in all sectors except heavy industry, which was intentionally constrained during a readjustment.

Regardless of the reality that the reform agenda was successful in several ways, it also yielded many significant problems for the country. The most notable of issues were concerning the environment.

From an economic perspective, China has been experiencing a 10% rise in national gross domestic product each year on an average over the last decade and then some. This has naturally contributed to excessive emissions. Data from *Council on Foreign Relations* shows that China has been responsible for releasing more harmful greenhouse gases like methane, nitrous oxide, and carbon dioxide than any other country in the rest of the world. Climate Watch notes that it crossed the United States as the top emitter of greenhouse gases in 2005 (Maizland, 2021).

Notably, China was also home to several of the most polluting cities in the World. Regional air pollution in the Beijing-Tianjin and Hebei provinces was terrible in 2014, with an annual average PM2.5 concentration of 93 micrograms/cubic meter (g/m3) considerably above Chinese national limits and WHO recommendations (WHO). China's Academy for Environmental Planning (CAEP) estimated in 2015 that emissions of fine particulate matter (PM2.5), Sulphur dioxide (SO2), and nitrogen oxides (NOx) significantly outstripped the cities' environmental absorption capacity by an estimated 80%, 50%, and 70%, respectively.

Now, *why is fine particulate matter (PM2.5) so dangerous*? Particulate Matter (PM) is a term used to describe a combination of solid particles and liquid droplets that are suspended in the atmosphere. These are divided into three categories: coarse, fine, and ultrafine. Coarse particles have a diameter ranging from 2.5 micrometers to 10 micrometers (about 25 to 100 times finer than a strand of a human hair), are comparatively heavier than fine particles, and as a result, tend to settle in the environment. Dust, spores, and pollen are examples of airborne particles. PM2.5 refers to particles with a diameter smaller than 2.5 micrometers (upwards of a hundred times finer than that of a human hair), which can stay suspended for an extended time in the atmosphere.

As a consequence of the combustion of fossil fuels and the chemical processes that occur in the atmosphere, these particles are produced. Nature-based activities like forest fires contribute to the formation of PM2.5 in the atmosphere. These particulates are also the leading cause of smog, which occurs when these particles combine. Contact to PM2.5 has a variety of short- and long-term health effects on the body. Coughing, sneezing, and difficulty breathing are all symptoms that occur in the short term, including irritation of the eyes, nose, and throat. Long-term exposure to PM2.5 may result in long-term respiratory issues such as asthma, chronic obstructive bronchitis, and heart disease, amongst other things.

While PM2.5 influences everyone, individuals with respiratory and heart conditions, children, and the aged are the most vulnerable to its effects. Because of its widespread prevalence, ambient particulate matter is a more potent killer than both alcohol and diabetes put together. An increase of 10 ug/m3 in PM 2.5 raises the overall mortality rate by 3–26%, increases the risk of severe asthma by 16%, lung illness by 36%, and heart attacks by 44% (King, 2019).

The State Environmental Protection Administration (SEPA) reported in 2006 that the woes of environmental ruination and the consequent rise in pollution levels had not come cheap. It had cost the Chinese economy what can be considered an annual equivalent of 10% of the Gross domestic product. They were not the only ones to come forward with such alarming claims. The Chinese media channels also reported that the economy had lost about thirty-six billion dollars in industrial production output because of the shortage of clean water needed to operate factories. The press has also reported a loss of thirteen billion dollars from corrosion of infrastructure and negative impacts on human health from acid rain. A loss of six billion from the growth of desert areas—which we can presume has resulted from severe dehydration and deforestation, which means that the land did not have enough water to retain its natural vegetative cover—to name just a few.

As a result of acid rain, one-quarter of the total land area and one-third of the arable land was damaged, with systems being eroded and reduced agricultural production. Air pollution in China was more than double the threshold deemed reasonably acceptable by the World Health Organization for suspended particulate matter concentrations. National death rates from chronic respiratory diseases were more than four times higher than in the United States. Asthma had reached alarming proportions in terms of harming children and youth.

Unavailability of adequate potable water, particularly in northern China, presented growing dangers to the country's economic development and its ability to maintain increasing urbanization levels. According to international standards, severe water scarcity is defined as the supply of only two thousand cubic meters or less of water per capita annually; 1,000 cubic meters per year is regarded as the bare minimum essential for existence. At the time, the average quantity of accessible water per capita per year in North China, which included almost 40% of the country's total population, was just one thousand and a hundred cubic meters per year. In the Beijing metropolitan region, the average fell below five hundred. Surface water contaminated to the point of being unsuitable for personal, commercial, or agronomic purposes accounted for about 40% of the total accessible amount.

As China's environmental challenge grew in magnitude, the country's share in global environmental issues increased in proportion. In addition to being the major importer of unlawfully logged wood, the NGO Global Witness identified China as the world's largest marine environmental threat in the Pacific. The World Bank identified it as a leading contributor to ozone depletion. China was the world's second-largest contributor to global climate change after the United States. China's economic growth significantly increased the country's pollution outputs. By the end of the first decade of the twenty-first century, China had surpassed the United States as the world's biggest user of energy and coal, as well as the world's largest car market and the world's highest emitter of carbon dioxide and sulfur dioxide.

Can you imagine how debilitating this is for a country's political system? Over four hundred thousand people have perished every year because of China's propensity to air pollution—and this is an expert opinion sourced from an environmental specialist at the University of Manitoba. The same specialist also estimates that about a hundred and ninety million people only have access to water that is so polluted and poisonous that they have fallen sick from resorting to drinking it out of the dire need and desperation to keep themselves alive. About forty million have migrated because their local ecological structures have not sustained them any longer. Don't think for a second that the public has not responded to this mammoth crisis—even with internal restrictions. Domestic media noted fifty thousand cases of environment-related protests in 2005 alone, some with an engagement of over thirty thousand people.

While problems became increasingly entrenched in the environment and impacted the political system through protests, the government was initially unwilling to look for alternative approaches. This is not to say that China did not have any environmental laws in place—these laws had been in existence since the 1970s. They included the Environment Protection Law of 1979, the Air Pollution Prevention and Control Law of 1987, and the Cleaner Production Promotion Law of 2002.

The first legislation even included provisions stating that the state must integrate state-formulated environmental protection plans into nationwide economic and social development plans. The state must implement economic and technical policies and measures beneficial to environmental protection in order to connect environmental protection activities with economic construction and social development.

The state was to enhance the growth of education in environmental protection science, enhance the research and development of environmental protection science and technology, and increase the scientific and technological standards in environmental protection. There were also provisions to improve the technical awareness of environmental protection while also making scientific understanding of environmental protection more widely known. Environmental protection would be the responsibility of all units and people.

Following the State Council's directives, the authorized agency of protection of the environment was deemed responsible for the integrated coordination and control of environmental protection activities across the nation. Environment protection administration departments of local people's administrations at or above the provincial level were responsible for conducting cohesive governance and monitoring of environmental protection activities within the jurisdiction of their respective local people's governments.

This law was followed closely by a series of other regulations that founded an ecologic legislative framework for China that seems press-oriented, focused on limiting the release of harmful substances rather than their conception, and wholly dependent on organizational monitoring and enforcement standards. Disparate regulations dealt with polluting sources such as water, air, solid and hazardous wastes, and noise. As initially drafted, these regulations required the creation of surrounding atmosphere, water, and noise standards, as well as concentration-based discharge limitations for pollution of both the air and water and decibel restrictions for noise pollution.

Hazardous releases by an operational facility that exceed the relevant limitations could "in-paper" result in administrative penalties against the facility owner. Although in most cases, the facility owner would be ordered to decrease the number of total emissions within a specific time. Solid and hazardous waste regulations impose fewer responsibilities than other environmental regulations, mainly due to the absence of efficient alternatives for the off-site disposal of such garbage during the early stages of the environmental legislative process.

As a matter of fact, the 1987 law that was made for protecting China's air and preventing air pollution from skyrocketing also had valuable features. These included:

- Establishing guidelines for reducing pollution
- Units that generate more pollutants than authorized by emissions requirements have the potential to be penalized and charged accordingly
- Putting in place an environmental surveillance system
- Addressing air pollution via the use of financial and strategic strategies
- Initiation and encouragement of investigations towards the care and mitigation of air pollution
- Forest-planting programs and initiatives to develop green urban and rural regions
- Environmental protection via the establishment of construction standards

The government tightened the criteria by revising the law in 1995, 2000, and 2014 to bring pollution below certain levels.

The Cleaner Production Promotion Law was passed in 2002 to promote eco-efficiency, cleaner production, increase the efficiency of resource provisioning rates. The law also served to mitigate and prevent unnecessary pollution of the environment, maintain, and improve the quality of the environment. The law aimed to protect and improve human health and promote long-term economic and social development. It had a lot of reasonable provisions—for instance, it stated that business elements, individual people, and corresponding management agencies engaged in manufacturing or providing services must develop and implement mechanisms for environmentally sound practices. These practices must follow the provisions for cleaner production as stated in this Law while functioning in the People's Republic of China.

According to the tenth five-year plan (2001-2005), the country's forest coverage would be increased from 11.8 to 18.2%, and the rate of the green area would be increased from 35% to 38%. The total quantity of leading agricultural and non-agricultural pollutants released was to be decreased by 10% compared to 2000, and additional measures were to be implemented to preserve and conserve natural resources.

Additionally, China participated in several international treaties and executed over a hundred policies and regulations surrounding the topic of environmental conservation. It didn't help—at least, not initially—because these rules fell essentially flat in terms of enforcement and when faced against the Chinese government's need to develop economically and become self-sufficient.

The nation was experiencing rapid economic development and was unwilling to sacrifice it for the sake of the employment, necessitating restrictions on its emissions, thus slowing industrial growth and reducing employment.

Making prosperity a reality appeases the people, rewards well-connected politicians, and delays future political reforms. Furthermore, the administration was rightly concerned that a significant slowdown would stir social discontent, alienate economic interests, and endanger the party's ability to maintain control. Aside from local governments, the central government also opposed the sorts of regulations, tax policies, and economy-oriented incentives for environmental conservation that have been implemented in other parts of the world. Even for the most polluting companies, land, water, power, oil, and bank loans were relatively cheap compared to other commodities. As a result, since central government language and actions remained elusive, there was a lack of enforcement from the local level of government.

Local officials would frequently disregard environmental regulations, assisting in reopening mines and industries that central authorities had shuttered.

But this is a book in which we discuss how China is doing so well, right? Then why all this?

Because, dear reader, you need to understand that should one be willing to bring about constructive change, nothing can stop them.

CHAPTER SUMMARY

- The Chinese government declared war on air pollution and implemented a comprehensive government effort to combat it because of the harmful effects on human health. However, to achieve the government's objectives, a significant amount of environmental investment in funds was needed.
- For nearly 4,000 years, China was a country where farmers were the center of attraction.

The irrigation infrastructure systems in China were already in place when the People's Republic of China was founded in 1949. Nearly all the arable land had been used for agriculture.

- The industrial sector has attracted the overwhelming bulk of Chinese investments and administrative resources since 1949, when the People's Republic of China was established. When China's authorities understood that the agricultural sector was hindering overall economic development, they focused attention on it.
- By diverting employees away from food production during China's Great Leap Forward, home blast furnaces and other businesses that did not support China's agricultural dependence caused more harm than good.
- As of 1978, China's socioeconomic structure started to shift toward a more free-market orientation. Agriculture, research, and military were all included in the scope of modernization, not only industry and technology.
- For most of Deng Xiaoping's presidency, the Chinese government prioritized reform and

openness as the country's basic policy, implying upgrading its economy and opening up to foreign trade at the same time.

- A decade and a half of economic growth in China has increased national gross domestic product of around 10% annually. As a result, there are a lot more emissions now.
- The emissions are created due to fossil fuel combustion and the chemical reactions taking place in the atmosphere. PM2.5 is formed in the atmosphere because of outdoor activities like forest fires.
- According to the State Environmental Protection Administration (SEPA), environmental ruination and the resulting increase in pollution costs the Chinese economy 10% of its yearly GDP, according to the agency's study released in 2006.
- There were mounting threats to China's economic growth and even it's capacity to sustain rising levels of industrialization, due to a lack of readily available drinkable water, especially in northern China.
- Enforcement was weak despite many conservation regulations, and the Chinese government's desire to grow commercially and become self-sufficient overrode them all.

- The country's economy was booming, and it didn't want to slow it down for the sake of the environment, which would have required limiting emissions, which would have slowed industrial expansion and reduced jobs.

In chapter two, we will read about the turning points, and the factors that led China to finally accept and work on a more environmentally conscious frame of development.

2

THE TURNING POINTS

Here's a simple fact: in the history of civilization and urban development, every nation that houses the ambition of becoming an industrial powerhouse cannot hope to achieve this without causing significant damage and havoc to the environment. From this perspective, if you look at the immense prowess and scope with which China has developed its finances and economy, you will note that the pollution that has resulted from all of it has no parallels.

According to a 2007 New York Times article, environmental pollution and deterioration were so deeply ingrained and had such far-reaching domestic and international ramifications that the pollution issue had gone beyond simply being a problem for Chinese civilians. It became an executive and legislative challenge that would undermine the legitimacy of the communist regime in power in the country (Kahn & Yardley, 2007). The pollution as of 2007 was so deep and horrid that it had made cancer the major contributor to the country's mortality rates. Atmospheric air pollution itself accounted for thousands of deaths annually, and over five hundred million people had no access to consumable water.

Here's a stark comment from one of China's top environmental scientists: *"It is a very awkward situation for the country because our greatest achievement is also our biggest burden"* (Kahn & Yardley, 2007). He goes on to note that while there had been pressure for change, it just hadn't done the trick. Worse still, it wasn't just about China anymore—the country's environmental issues have repeatedly hurt the world at large. As witnessed, sulfur dioxide and nitrogen oxides resulting from China's heavily burdened coal-combustion-based industrial powerhouses fell as acid rain on cities like South Korea's Seoul and Japan's Tokyo. The Journal of Geophysical Research has even noted that particulate pollutant matter all the way from China has found its way into Los Angeles.

For decades, China held out on making any commitments to the United Nations framework related to environmental conservation. According to high-level

Chinese politicians, the brunt of environmental pollution issues should have been addressed by developed countries like the United States, which enjoyed the easy growth of their already burgeoning economy with few (if any) limitations. The bothersome thing is that while the rest of the world had turned its attention to environmental issues like global warming, China was still focused on all policies surrounding its internal industrial revolution and developments.

Until China was forced to go down a route of introspection, the internal parties always considered that the outside world played a role in the deterioration of China's environment, which is genuinely not justifiable. Think about it: of course, environmental degradation is the total of a lot of things, but just because Person A and Person B are contributing to something together, does that mean that Person A gets to go on doing it while putting the brunt of all the blame on Person B? It doesn't work that way.

The logic of these political personalities is that due to manufacturing necessities, Chinese manufacturers are forced to pollute the environment by dumping pollutants into rivers or releasing bouts of poisonous smoke into the atmosphere to make cheap products for companies based in the United States and Europe. Western nations often contract labor from the East because the currency differences account for reasonable salaries for people in the East—making it a cheaper option for them by far. Truth be told, major powers like the US, Britain, and Japan did very poorly in environmental conservation and more or less destroyed their native environments on their journey to prosperity. They only turned their attention to the environment *after achieving a reasonable level of development and becoming established and economically stable.*

As a result, international investment in China increased as global firms expanded their operations in the country. Beijing also reiterated that it would not accept any mandated limitations on its carbon emissions, which would virtually undoubtedly hurt the country's rapid industrialization. They validated the logic of the argument, which was that rich countries were responsible for climate change and should find a method to address the problem without interfering with China's development.

This was essentially the principle of "*common, but differentiated*" responsibility division—and a number of developing nations used it to elicit exemptions from their role in emissions leading to environmental pollution and degradation. According to Ren Yong, a seasoned specialist at the Center for Environment and Economy, based in Beijing, industrialized countries mainly deal with environmental problems *once they have become rich.* But developing countries like China have to deal with them while they are still poor. As a result they have *no model to rely on, and neither do they have any assurance that there will be no public outrage over halted economic progress.*

So, for obvious reasons, China's momentous economic progress, which came hand-in-hand with excessive dependence on fossil fuel and lack of penalties for polluting, went on hampering its environment. By the end of 2010, China was already the largest consumer of fossil fuels like coal and the energy derived from burning it. It was also the biggest locomotive market and the most prolific emitter of greenhouse gases like carbon and sulfur dioxide. There was a time when it was known as *"airpocalypse,"* a term given to the condition of China's environment by the international news platforms.

Several studies have repeatedly shown that air pollution may cause a substantial rise in premature mortality in China. A growing body of evidence from epidemiological research since the 1980s indicates that bad air quality in northern Chinese towns is associated with severe health consequences, including respiratory, cardiovascular, and cerebrovascular disorders.

It was also shown in medical and epidemiological studies that air pollution (especially particle pollution) may cause brain-based inflammatory responses, affecting brain function and aggravating depression and anxiety (Greenstone et al., 2021). It would be possible that air pollution would have an impact on people's intellect and judgment processes. According to several investigations that look at a broad range of experimental situations, premature mortality in China may be substantially aggravated by air pollution. When looking at China's acid rain regulations, Tanaka (2015) discovered a substantial reduction in infantile mortality in zones with stringent regulations on SO2 emissions. In a study in 2016, He and colleagues estimated that a 10% reduction in PM10 emissions, as measured during the 2008 Beijing Olympics, results in an 8% reduction in the monthly average mortality rate. Bombardini and Li (2020) study trade-induced harmful emissions and find that China's infant mortality rises by 4.1 fatalities per thousand live births when pollution concentrations rise by one standard deviation.

The Chinese people became incensed by severe pollution occurrences in big metropolitan areas like Beijing and other major cities. In reaction to the lack of basic public information about pollution and the absence of governmental measures to solve the issue, an outpouring of criticism ensued. Under these circumstances, Chinese policymakers started deviating from their previous approach of putting economic development ahead of environmental issues. Premier Li Keqiang launched a war on pollution at the beginning of the annual conference of the People's Congress 2014 in March 2014, asserting that smog was nature's warning against wasteful and blind growth and calling for a halt to it.

This has to be clarified: China's air pollution levels remain high, indicating that more measures are needed to improve the country's environmental quality in accordance with international standards. To give you an idea, in 2018, the country's yearly average PM2.5 concentration was 40.1 g/m3, which is still four times more than what the World Health Organization (WHO) considers reasonable.

However, to the government's credit, it did not just sit back and wait for the consequences of its pronouncements. It implemented significant regulatory reforms on several fronts.

PM2.5 was perceived as a significant pollutant, and national maximum limits for PM2.5 were established for the first time; a national, real-time air controls systems and disclosure operation was launched. Pollution reduction was set as a bureaucratic objective in government officials' assessment and promotion.

Various environmental measures were implemented, such as piloting seven regional CO2 cap-and-trade systems and encouraging electric vehicles (EVs).

Chinese authorities have made considerable significant efforts to improve air quality since 2014. The amount of pollution in the air has decreased considerably in the local area. Five years after the high in 2013, PM2.5 levels at the national level had fallen by about 40%. While at the same time, SO2 and CO concentrations had decreased by 65% and 33%, correspondingly. In contrast, after adopting the Clean Air Act, it took the United States a minimum of a decade and two major economic recessions to achieve a similar proportion of deductions in air pollution.

More than a decade of strong economic development and tacit acceptance from other countries have helped Chinese political authorities maintain long-term sustainability. Chinese authorities are now very concerned about their worldwide image. They are implementing an advertising strategy that promotes China's advantages as an international economic partner while downplaying it's political and military might in the international arena. Because the Communist Party benefits from the Western assumption that economic development would eventually lead to more political freedom, public statements of worry from

Western government and corporate executives about China's failures are muted, convenient for the Communist Party. China's influential position as a factory to the world (since so much of the products distributed worldwide owe their origins and parts to China) is backed up by a strong track record of performance. In the eyes of the international business world, Beijing has made continuous infrastructural development, taken measures to make state-owned companies more competitive, and reawakened the long-suppressed ambition of many Chinese to establish their own businesses. In the last two decades, Chinese manufacturers have established a global reputation for producing and delivering low-cost, dependable goods in a short amount of time. As a result of these positive connections, China's image has been more disassociated from major negative associations such as those associated with the ideologies of Mao Zedong and the Communist Regime, the Cultural Revolution, and the events of 1989 Tiananmen Square.

China's remarkable economic development has been helped by its ability to project a positive national image abroad. It has also lessened popular momentum in the West to take revenge against China over growing inequality or to push for political, ecologic, or human rights changes in the country, among other things. However, the widespread recognition of 'Brand China' has both negative and positive consequences. Product recalls are threatening to damage the reputations of all Chinese manufacturers.

Because of the 2008 Olympics, China drew the world's attention to its accomplishments and its faults. Although the Communist Party does not want to relinquish political power and local officials seem to resist corrupt practices reforms, Beijing has a more favorable foreign image than it had twenty years ago. China obviously wants to be a major global participant in many spheres of life, including economics, politics, and culture.

On the other hand, China cannot but attract more and more attention as its influence grows. The desire of other global powers for stability in the world's most populous nation and their growing reliance on China to maintain inflows of inexpensive consumer products may serve as a buffer against demands to diversify and change political and economic processes in the country. Even yet, China cannot afford to ignore the importance of managing its image. For starters, the nation now stands to lose more in terms of economic prosperity. China understands that to increase its political and economic influence; it must satisfy or seem to live up to the standards of positive behavior set by international bodies like the United Nations, the World Trade Organization, and the Organization for Economic Cooperation and Development.

Until the government decided that it was high time to turn their attention to the problems of the environment, China's air was chock-full of pollutants. These pollutants resulted from emissions of industries, power plants, fossil fuels burnt by cars, and the harmful chemicals produced by the construction industry. Particulate matter constituted a grave threat to citizens, with smoke, dirt, chemicals, and dust poised to cause severe negative ramifications. Let us look at this instance: in 2012, a study was published by the US Centers for Disease Control and Prevention. They measured the air quality of a smoking lounge in an American airport and noted that the average particulate matter rating was 166.6 micrograms every cubic meter. In 2013, the daily average particulate matter reading in Beijing was 194 micrograms per cubic meter. This means that the quality of breathable air in Beijing was worse than what could be found in a smoker's lounge! At least, with the smoking lounge, you have the option not to enter it and harm yourself. If, however, you happen to be an inhabitant in a city, that is hardly an option for you.

Between 1981-2001, a research team from MIT (Massachusetts Institute of Technology) and Tsinghua and Hebrew universities found that the average quantities of particulate matter in North China were more than 50% higher than that of South China, probably owing to the heavy dependence of the north on power generated from coal industries. This implies that the life expectancy of people in North China is also at increased risk. Indeed, the study found that the five hundred million people living there lived an average of 5.5 lesser years than the residents of South China. The cause? *Respiratory illness.* We can shun the idea of how dangerous environmental pollution is as much as we want, but the data cannot be disproven. When there is too much effort on ruthlessly building the economy with no concern for nature, it does strike back, and unfortunately, the brunt of the burden also falls back on us (Chen et al., 2010).

People who breathe China's air are clearly suffering grave health consequences. This is not the only shocking study to exist out there. In 2010, the Global Burden of Disease Study found that the particulate matter present in outdoor air pollution was alone responsible for more than a million premature mortalities in that year. Alarmingly, the study also discovered an increase in the numbers of cardiovascular ailments in China, such as ischemic heart illnesses, pulmonary strokes, and heart failures, all increasingly linked to polluted air (*Global Burden of Disease Study 2010*, 2012).

However, things slowly began to change with the Beijing Olympics in 2001. After winning the bid to host the Olympics, Beijing began to make substantial efforts to invest in environmentally friendly construction tools. China also turned towards using sustainable energy to live up to the international image that would now be so important as they would be hosting one of the most critical games in the history of sports. Following the 2008 bid, people commonly thought that Beijing's air quality was just as bad as that of Mexico—Beijing and Mexico City were tied to being the most polluted capitals to exist. However, Beijing outbid Toronto and Paris to become the city that would host the 2008 Olympics because they vouched to go green.

At the time, the Beijing Games risked being labeled as the *Smoglymipics.* Beijing made significant investments in environmentally friendly construction raw materials and sustainable sources for the Olympic Village. The city also made a conscious effort to decontaminate the city's notoriously polluted air and water: two hundred polluting businesses in Beijing were shut down or renovated, and Beijing would remove more than a million vehicles from city streets for the duration of the Games. Following Beijing's ambitious and vigorous campaign against environmental pollution, blue sky days with the lowest levels of air pollution increased to

274 in 2007, an increase from around 180 in the previous year.

Even though the efforts to clean the air were previously anticipated to last only during the lead-up to and organizing of the Summer Olympics, UNEP authorities confirmed that the significant gains made had created awareness of and demand for improvement in quality of life among the inhabitants of Beijing and municipal authorities going forward. Several efforts under the Blue Sky Program included upgrading or shutting almost 200 local businesses, converting cars and buses to run on cleaner fuels, and promoting clean transportation pilot projects. The Olympic Village was one of these businesses. Clean transportation efforts were regarded as a priority in Beijing, which is at times too cosmopolitan for its own good—given the sheer number of vehicles that its citizens are constantly reliant upon. In addition, city authorities increased the volume of open vegetation and green spaces available for people to appreciate long after the Olympics were over.

According to estimates, more than thirty million trees and shrubs were planted in the city's heart, resulting in the establishment of about seven hundred and twenty green areas. Numerous restrictions, such as those on highly polluting truck traffic, were kept in place for months after the games were over. According to the United Nations' evaluation, the 2008 Olympic Games made significant advances in many other categories. Municipal solid waste recycling objectives were surpassed by 5% while recycling at Olympic venues exceeded the target rate by 23%. Although coal still accounts for more than 40% of Beijing's power usage, renewable energy sources provided about 20% of the energy required to host the Games.

In recent years, demonstrations against environmental injustice have grown in frequency and size, as have environmental non-governmental organizations (NGOs). Opposition against coal-fired power stations, garbage incinerators, chemical industries, oil refineries, and other projects has drawn large crowds to the streets. Environmental protests have increased by an average of 29% a year since 1996, according to Yang Zhaofei, vice-chair of the Chinese Society for Environmental Sciences. In 2011, environmental demonstrations increased by 120%. A prominent senior figure of the Communist Party's policy and Legislation Affairs Committee is of the opinion that pollution, rather than territorial disputes, is the primary driver of societal unhappiness in China.

Concern for the environment is becoming more widely accepted. A Pew Research Center survey in 2013 found that half of the Chinese respondents said air pollution was a very severe problem, up 11% from the year before and up 16% from the year before that. Social media sites are increasingly focusing heavily on pollution. There is even news that a man in Beijing's bordering Hebei province sought compensation for expenditures linked to the city's smoggy air by purchasing filtration systems devices, face shields, and workout equipment for use inside. Despite the lawsuit's failure, it got positive coverage in the state-run media (Gardner, 2014).

When a chemical factory in Zhangzhou, Fujian region, generating paraxylene (often referred to as PX and employed in producing polyester and plastics) exploded in April 2014, protesters' rage was validated. As a result of the blast, the Chinese army's anti-chemical defense unit was activated, and 30,000 people were evacuated. Violent anti-PX protests on a large scale have occurred in the last eight years in Guangdong and Yunnan provinces in the south, Zhejiang, and Fujian in the east, and Liaoning in the north, as well as Sichuan province in the west. An anti-PX campaign launched in 2007 has brought the problem full-circle, and the latest tragedy in Fujian illustrates the government's limited options. The options available to the government are highly constrained. Most individuals are concerned about pollution as a major issue. They will be dealing with this issue as long as there are reminders in the water and air they breathe. After seeing a change in public opinion, the central government reacted to concerns and discontent. At the National People's Congress (NPC) in March 2014, President Xi Jinping said that the administration would act with an iron fist at any offenders who impair China's ecological balance or environment.

A significant worry for Chinese President Xi Jinping's government has been climate change. Beijing has taken several steps to address this issue. As per his estimates, renewable energy sources should account for 25% of total energy usage by 2030. China is geared towards achieving carbon neutrality by 2060, achieving maximum greenhouse gas emissions before 2030, and lowering the overall intensity of carbon emissions, or the carbon content emitted per unit of gross domestic product by more than 65% by 2030. Efforts are also being made towards installing enough solar and wind energy generators to have a total capacity exceeds more than one billion kilowatts by 2030. There is an increased impetus for increasing forest coverage by six billion cubic meters by 2030.

CHAPTER SUMMARY

- For a long time, China's leaders believed that the outside world was to blame for the degradation of China's environment. This was a mistake that cannot be justified.
- People, both within and globally, continued to protest about the condition of China's environment for years.
- Things came to a turning point from the Beijing Olympics because China did not want to earn the reputation of being the most polluted host globally.
- Today, activism and people's protests have come a long way in helping the battle to improve the environment.
- In March of 2014, China's president, Xi Jinping, promised the NPC that the government would deal harshly with anybody found guilty of harming China's environmental integrity.

In the next chapter, we will read about China's environmental goals and what it has achieved in this time.

3

CHINA'S ENVIRONMENTAL GOALS

There have been efforts to enhance the city's air quality since the Summer Olympics in Beijing back in '08. Beijing's strict environmental regulations led to a 30% improvement in air quality within just one year, resulting in a proportional decrease in cardiopulmonary and respiratory illness.

In 2013, China began a campaign to improve the air quality in its urban areas. Science had discovered the causes of environmental pollution and hazardous rain; these included emissions from automobiles and factories, hay burning on farmland, and other sources. Increasingly stringent regulations and stricter enforcement led to the relocation of several industries while forcing others to shut down entirely because of the pollution they produced. The government provided farmers with incentives so they could recycle hay rather than burn it as fuel.

However, since China is highly reliant on manufacturing, the costs of halting production and moving or closing down businesses may be prohibitively expensive. Similarly, many developing countries have been confronted with the challenge of striking a balance between revenue, economic expansion, and the well-being of their population densities.

To address this problem, China is redirecting additional infrastructure investment towards renewable energy technology. In 2016, coal accounted for roughly 60% of China's energy mix, compared to just 5% of solar power, according to the International Energy Agency statistics. However, things are changing.

To its credit, China decided to take the concerns surrounding environmental pollution seriously. We just studied that China's heavy-duty economic development plans and programs led to severe degradation of the country's environment. The international news media and the people equally denounced the hazardous health conditions imposed by environmental degradation. There was an outrage over the absence of essential information related to how polluted the environment had become. Similarly, an equal distrust of the government persisted owing to its inability to take a stand.

When Li Keqiang and Xi Jinping declared war on pollution in the People's Congress of 2014, China was already moving away from ruthless economical growth at the cost of further destruction of the environment.

IMPORTANT LAWS

An essential energy source for most of China's development has been coal, which is both inexpensive and environmentally harmful. In 1990, coal accounted for more than 70% of the energy used in China. However, China has not been one to back down and has come up with a number of important laws. Among these laws is the 2013 *Airborne Pollution Prevention and Control Action Plan.* The issue of regional air quality, which is distinguished by fine respirable particulates and particulate matter, has grown more significant, affecting people's health as well as social peace and stability, has become increasingly apparent. As the urban expansion progressed, the consumption of energy resources increased, and the need for air pollution management became more pressing. In order to enhance the quality of the air in China, the government put this strategic plan together.

The plan acknowledged that air pollution was a significant problem in China. It identified coal as an essential source of air pollution, imposing limits on its use and limiting its emissions. A primary goal of the strategy was to decrease pollution while also improving air quality in critical areas. Other objectives were to reduce urban Particulate Matters (PM10) concentrations by 10% by 2017 as compared to 2012 and to raise the number of days with reasonably acceptable air quality on an annual basis in the process. The idea was to do this by altering the energy structures and investing in new energy sources, among other things.

This system would see the establishment of new prevention and control of air pollution mechanism in which the government would play a leadership role while also integrating enterprise-led efforts, market forces, and public involvement.

During the Thirteenth Five-Year Plan period, China published several plans to reduce PM2.5 levels in China's ten worst polluted cities by 18% and reduce coal consumption to 58% by 2020. China released these plans in November 2016. To significantly reduce severe pollution over the time covered by the Fourteenth Five-Year Plan, the Communist Party's Central Committee proposed in November 2020 that particle matter and ozone regulations be tightened.

In 2015, the *Environment Protection Act* was introduced. Businesses were obliged to publish their pollution levels, as well as the number of penalties they had paid, on a public website. Under the new legislation, companies that release pollution levels greater than approved limits would be subject to daily penalties (an increase over the previous regulation passed in 1989). Public officials were to be evaluated based on their progress in achieving their environmental protection objectives. Citizens now can report ecologically harmful actions carried out by businesses or individuals—*and* the ability to report the failure of environmental protection regulatory authorities to take action when necessary.

The People's Republic of China, along with the European Union and several other nations, is attempting to establish an emissions trading strategy, which would require polluting offenders to pay for environmental damage and, as a result, encourage them to decrease their emissions. It will first concentrate on coal-based power and gas-fired power facilities. China's attempts to transition away from coal and toward renewable energy are important, and the nation has already achieved some strides in this direction. Compared to a decade earlier, alternative energy sources currently account for almost 15% of China's energy supply (as of 2019).

The Chinese have been using hydropower for many years. As the world's top producer of solar and wind power, the country increases the number of solar farms, solar panels, and wind turbines installed. As part of this expansion, it is building seventeen nuclear power reactors, operational by the middle of 2021. Furthermore, the usage of electric vehicles is being encouraged in Beijing and other regions. According to the Chinese government, 1.37 million cars running on renewable resources (including rechargeable electrical and hybrid-electric cars) would be marketed in China in 2020, representing a roughly 11% increase over the prior year.

China has also participated in regular conferences with conservation officials from Japan and Korea and formed climate accords with India, the world's third-largest producer of greenhouse gas emissions. During the Obama administration, the cooperation between Chinese and American businesses, scientists, and specialists on renewable emissions and energy capture technology has grown significantly.

China has also committed itself to tackle global climate change via the 2015 Paris Agreement. Part of the agreement required China to attain net-zero carbon dioxide emissions by 2030 and achieve a 20% share of total consumption from non-fossil fuels within that time frame. The Chinese government has also committed to having renewable technologies account for 25% of total energy usage by 2030. They are reducing carbon concentration (the amount of carbon released into the atmosphere per unit of GDP) by more than 65% by 2030 and implementing enough solar and wind energy turbines to have a total capacity of 1.2 billion kilowatts by 2030.

For years, China has reaped enormous benefits in know-how and funding from the Clean Development Mechanism. The United Nations emissions trading system allows developed nations with Kyoto Protocol commitments to reduce emissions to implement emission-reduction initiatives in developing countries.

China has progressively risen to the top of the global rankings in developing renewable energy sources and the emissions trading industry. China has surpassed the United States as the world leader in renewable energy generation and associated equipment manufacturing. Its share of worldwide wind-generated energy production and photovoltaic panel manufacturing increased to 45% and 72%, respectively, from what it was in 2020.

Between 2010 and 2020, China's installed wind capacity grew nine times, from 31 to 280 gigawatts, while its installed solar capacity increased six times, from 42 to 250 GW, over the same period. Since 2010, the costs of solar, wind, and rechargeable batteries have decreased significantly, and they are likely to continue to decline, making it feasible for China to persist in substituting coal-fired power generation with renewable energy sources on a large scale in the foreseeable future. The Belt and Road Initiative (BRI) plan encourages Chinese companies to build renewable energy projects in underdeveloped countries (Maizland, 2021).

The Belt and Road Initiative (BRI) focuses on major infrastructure expansion, economic cooperation, and trade and development is China's most visible card in third-world foreign relations and legislative diplomacy. More investment will be made in climate adaptation and mitigation infrastructure when the Belt and Road Initiative is coupled with China's environmental efforts. This includes supply of water, public health, coastal protection, and solar and wind energy facilities.

Chinese investment in renewable energy and other eco-friendly technology is anticipated to rise in the years to come. It is projected that the country's carbon neutrality strategy would need an expenditure of about fifteen trillion dollars over the next thirty years. The government has also committed to increasing forest coverage, with an estimated six billion cubic meters of forest coverage by 2030. Additionally, President Xi Jinping stated in 2020, through a video conference to the United Nations General Assembly in New York, that the country's goal is to reach peak emissions by 2030 and carbon neutrality by 2060.

While this is a lofty ambition, there is hope that China will rise to the test. It has already undertaken several measures that have benefited the environment. The following three chapters will cover some of these measures in detail.

CHAPTER SUMMARY

- Even before the People's Congress formal declaration on polluting in 2014, China had already started shifting away from unrestrained economic growth at the cost of growing environmental destruction.
- China's infrastructure investment is moving fossil fuels to renewable energy.
- China's 2013 Airborne Pollution Prevention and Control Action Plan singled out coal as a significant source to fight smog, imposing severe limits on coal use and lowering coal emissions.
- During the Thirteenth Five-Year Plan period, China proposed several measures to reduce PM2.5 levels in China's ten most polluted cities by 18% and coal consumption by 58% by 2020.
- As a consequence of this predicament, the Environmental Protection Act was enacted in 2015. Businesses were obliged to publish data

about their pollutant emissions and penalties paid on a public website.

- The People's Republic of China, the European Union, and many other nations are trying to develop an emissions trading system to incentivize polluters to pay for their environmental damage and, as a result, decrease their emissions.
- The Chinese have depended on hydropower for a long time. The amount of solar farms, solar panels, and wind turbines getting constructed in the world's leading solar and wind power producer continues to grow.
- President Xi Jinping said in a teleconference to the United Nations General Assembly in New York City in 2020 that China intends to reach peak emissions in 2030 and carbon neutrality in 2060.

4

COLLECTION AND ACCESS TO INFORMATION

In this chapter, we will look at all aspects relating to environmental data and its disclosure, with the intent being on how this has resulted in long-term improvements in air quality and air pollution reduction.

China's fight against polluted air is part of a more considerable effort to address human health and the environmental catastrophes caused by rapid modernization in recent decades. In Tangshan, which used to be a city suffering in the aftermath of a devastating earthquake, economic growth has pulled hundreds of thousands of people out of financial hardship out of complete devastation. In exchange, many of them have been forced to live with undrinkable water, contaminated food, and polluted air. The result has been public pressure to disclose legitimate data on pollution.

Public pressure has been so high that the Beijing Municipal Environmental Protection Bureau (BMEPB) was forced to take the stand to publish general statistics regarding the quality of air in the city. The United States embassies have the policy of monitoring worldwide air pollution, and it also distributes this data in real-time.

This covers data collected from China. In 2008, the United States Embassy in Beijing started the country's first air-quality monitoring program. The air quality index (AQI) provides a numerical number ranging from one to five hundred and a qualitative evaluation of the environmental quality and a warning about potential health consequences. To create the AQI, the EPA combines data from five distinct pollutant concentration indices: particle pollution (PM2.5 and PM10), ozone (O3), monoxide (CO), hydrogen sulfide (SO2), and nitrogen dioxide (NO2). China has tightened its pollution laws and improved its air quality due to the embassy's public release of air-quality monitoring data and scientific contacts with the United States Environmental Protection Agency. It is a crucial way of keeping the citizens of a country aware of what is happening concerning environmental transparency and protecting themselves from air pollution.

The air quality index (AQI) is equivalent to the maximum values assigned to each pollutant. Everyone should limit their outside activities when the AQI rises to unhealthy levels over 150. According to the governments of both the United States and China, everybody should eliminate all outdoor activities when the AQI increases to dangerous levels above 300. Since 2008, the American Embassy in Beijing has tweeted hourly and daily particle counts, health-based evaluations of air quality, and other information.

Now, there is a chance that the official quality of air was misreported before 2013. A report published by Thomas Stoerk for the journal *Atmospheric Environment* in 2012 states that the official misrepresentation of environmental data available in China was due to the presence of a few observations that go above the *Blue Sky Day* threshold. The Central Government of China has always used this to rank air quality in Chinese cities.

The pattern appears to have been that China's contamination was reported deceptively, preventing the people from comprehending the situation and allowing the government to remain inactive. The pollution of outdoor air in China is believed to be responsible for twenty-five million valuable years of potential life squandered each year. According to experts, life expectancy is estimated to be reduced by three years in a severely polluted Chinese city, or roughly 12 days for every one-year exposure. It is estimated that residing in a moderate-to-average polluted Chinese city reduces one's life expectancy by almost two years, or roughly six days for every one year of exposure. According to the World Bank, all seventy-four Chinese cities evaluated in 2013 had yearly pollutant concentrations worse than "moderately polluted," and more than 60% had levels worse than "extremely polluted."

This threshold measures air pollution and is used to evaluate local officials. Suspiciously, there were a considerable number of observations that just fell short of this threshold. What happens is this: if local state governments can claim a high number of blue sky days, it becomes a way of appeasing the public. Because government officials react by reporting the required statistics incorrectly, the issue is that increasing political pressure to address air pollution may result in quantitative data manipulation and malpractice.

AQIs, or air quality indices, were intended to help the public better understand pollutant concentrations and their impact on human health. However, due to discrepancies in measuring and reporting techniques, the general public has been misinformed about pollution levels. The Environmental Protection Agency of the United States launched a comprehensive air quality evaluation system in 1976 due to the proliferation of various indicators. When public reporting started in China in 1997, the country used a comparable index, but the evaluations of measurements radically differed. For example, in the United States, levels designated as "unsafe" were called "light pollution" in China.

When "light pollution" is used in other countries, it usually refers to fewer pollution levels than the yearly norm. However, in China, the term "light" pollution refers to levels above three times the average annual standard—eleven times what the WHO classifies as safe yearly emission rates.

According to the qualitative composite score in the Air Quality Monitoring Index Analytical Regulations, the same moderately polluted cities can officially claim up to 365 days of outstanding or high air quality, which is highly controversial. The average pollution level in Beijing would be passable under this scoring, even though fine particle concentrations in the city are almost twice as high as in a highly polluted metropolis. People in China are eager to learn about environmental pollution's potential adverse health consequences and take precautionary measures to prevent exposure to harmful pollutants, such as donning a pollution shield to protect their health.

Despite the development of mobile applications and websites that provide information about the environment—three apps have been installed over a million times by the Chinese population—there is still a great deal of misunderstanding about what these and many other air quality evaluations imply.

There are a variety of factors contributing to the misunderstanding around these applications and their use. While many mobile apps and websites disclose concentrations of pollutants from the US Consulate and Chinese monitors based on the daily concentration levels of the US index system and evaluations, others use the Chinese index system and evaluations or a combination of the two. Now, these parameters are vastly different, and combining or comparing them can result in a lot of confusion. The English version employs US Environmental Protection Agency evaluations, while the Chinese version uses the Chinese assessments, even though they are significantly different. One of the most widely used applications makes the error of relying on the English version to disclose data—naturally resulting in discrepancies.

The Asian Development Bank (ADB) believes that its comprehensive Inclusive Green Growth Index (IGGI) would close this gap at the international level. According to the Asian Development Bank's statistics, Singapore, Bhutan, Sri Lanka, and Laos are among the countries in Asia that have achieved the best environmental sustainability results. Turkey, Uzbekistan, and Mongolia round out the bottom three. The International Greenhouse Gas Initiative (IGGI) evaluates nations' economic and social indicators and environmental records. With 28 indicators, the IGGI is more comprehensive than previous attempts of a similar kind. These indicators include clean water availability and levels of air pollution.

However, the good news is that while the work of earlier researchers suggests that the official data on China's air quality was subject to an amount of manipulation, recent data dating to 2020 indicates that the government can no longer manipulate evidence. Modern systems for monitoring and evaluation have digitized and automated the reporting and sampling procedures, which reduces the scope for human interference and data manipulation (Greenstone et al., 2021).

For example, here is a bit of real-time, honest reporting: according to official statistics, Beijing had moderate or better air pollution levels for 307 days in 2013. Even when pollution levels are fifteen times higher than the WHO annual guidelines, the Beijing government classifies the air quality as "moderate." Beijing's yearly respirable particle accumulation of 90.1 micrograms per cubic meter (and roughly 103 micrograms per cubic meter at the US Embassy) was more than nine times higher than the WHO guidelines. The research found twelve places with levels much worse than Beijing, notably Xingtai in Hebei province, which had an average intensity of 155 micrograms per cubic meter. Beijing was the worst of the cities studied.

Nationally, the Chinese government has made significant strides in improving its pollution statistics to drive its environmental policy, including the world's biggest carbon-pricing scheme, which covers seven provinces and is the world's largest in scale. Examples include high-frequency data gathered via Ongoing Emissions Monitoring Systems, which offer critical information on the origins of air pollutants, allowing the government to establish appropriate incentives for firms to reduce their emissions and reduce greenhouse gas emissions.

So, the positive news is that based on research, air quality data misrepresentation seems to have stopped after 2012 (Stoerk, 2015). Launched in 2013, the real-time environmental monitoring and disclosure initiative was designed to provide real-time information on pollution. It was a watershed moment in the availability and awareness of pollution information. More than sixteen hundred U.S. Environmental Protection Agency-grade monitoring stations were built by the government, each equipped with mechanized, legitimate surveillance equipment that tracked densities of six criterion contaminants. In less than two years, the security apparatus had successfully expanded to encompass the whole country.

The central government has developed a modern data streaming system in which findings from on-site monitoring are sent in real-time to city, provincial, and central-level government officials. Automation has dramatically improved the quality of China's air pollution data, and the patterns in post-automation data from real-time monitoring data are similar to those in satellite-derived pollution measurements,

According to Barwick et al. (2019), comprehensive data on a range of impacts, including pollution awareness, short- and long-term economic activity, and health repercussions, offer convincing proof of the program's overall effects. In a study based on an incident-design approach that considered a phased implementation of a pollution information campaign, the authors demonstrate that the plan altered perceptions of accessibility to pollutant information and significantly increased household awareness of pollution problems.

In response to these changes, short- and long-term patterns in household activities were triggered, including an increase in online search queries for pollution-related topics, modifications in daily consumer preferences to reduce exposure to pollution, and increased water treatment plans for households in less polluted areas, among other things. Due to the long-term and short-term behavior changes that resulted from the effort, China reduced the fatality cost of air pollution. They accomplished this by smoothing out the dose-response curve of pollution on mortality and, as a result, reducing the premature death cost of air pollution.

More recently, China's government has mandated that over fourteen thousand coal firms submit time-series analyses on pollutants from their units produced to a publicly accessible internet portal since the conclusion of 2013. The Chinese government used such information to guide the development of new laws and regulations to decrease emissions. As per a recent study, the new rules were responsible for a 13.9% reduction in Sulphur dioxide emissions from fossil-fuel-based power plants (Karplus et al., 2018). I should emphasize that this method is only effective if, as is the situation in China, businesses are subjected to severe fines for damaging the environment and providing false information. Data alone should not, however, decide policy. Aggregate results frequently reflect the impact of individual behavior, attitudes, and events since economic analysis is a social science. Environmental policymaking is no different: interventions may fail for various reasons that have nothing to do with their design. The point here is that there are efforts in the right direction—the desire to do better and keep the public in the know-how of what is happening. When a country moves in this direction, positive outcomes naturally follow.

CHAPTER SUMMARY

China's battle against dirty air is part of an overall endeavor to solve the environmental and human health disasters created by recent fast modernization.

- The United States Embassy in Beijing launched the nation's first air-quality monitoring system in 2008.
- The public pressure was so intense that it compelled the Beijing Municipal Environmental Protection Bureau (BMEPB) to

adopt the position of publishing public data on the city's air quality.

- China has tightened its pollution laws and improved its air quality due to the embassy's public release of air-quality monitoring data. It is a critical method of keeping a country's people informed about what is going on in ecological transparency and safeguarding themselves from environmental pollution.
- There is a possibility that the official air quality was misreported before 2013. This was due to local agents' desire to avoid penalties. Also, there was a lot of uncertainty regarding how environmental contamination would be reported through apps, mainly since they were based on ambiguous criteria.
- While previous research suggests that official data on China's air quality was subject to manipulation, recent data from 2020 indicates that it can no longer exploit documentation. This is because of modern monitoring and evaluation systems that have digitized and automated reporting and sampling procedures.

In the next chapter, you will learn how China is increasing access to clean energy through financing.

5

PROVISION OF AND ACCESS TO FUNDS

As part of its global initiative to show that it is genuinely working to reduce pollution, China has devoted a lot of funds over the consequent years, intending to ensure that there is enough finance to back the ecological problems it wants to tackle. Given that China has been trying to discover new methods to establish itself as a global leader for some time, it was a reasonable next step, especially given that ecological politics is a multibillion-dollar game to play. While China has been attempting to identify various financial streams for environmental improvement, it has encountered significant difficulty in finding funds to clean up the damaged land and water. However, that has not stopped it from making efforts to allocate funds.

The first environmental fund solely dedicated to ecological improvements raised eighty-eight billion yuan (the equivalent of almost thirteen billion dollars) during the first phase in 2019. The purpose of the fund was to increase investments in "green" environmentally sound projects and businesses.

The Chinese Ministry for Ecology and Environment director, Zou Shoumin, was responsible for handling investments. He unequivocally stated that the National Green Development Fund would be a key investment component in strategic nationwide programs to develop environmentally sound industrial practices. This included the development of the Yangtze river bank in an ecologically sound way. In the words of the environment minister, Huan Runqiu, fund allocation was key to initiating the "green transformation" of China's economy. It showed that the market had played a considerable role in reducing the harmful impacts of air pollution (Xu et al., 2020).

Between 2016-2020, the Central government in China has systematically allocated 97.4 billion yuan to stave off air pollution, 78.3 billion yuan to fight water contamination, and 28.5 billion yuan to improve the soil. Additionally, it has invested 20.6 billion yuan into fighting ecological issues in rural regions (Xu et al., 2020). The hope is that this initiative will continue. The

Chinese government will continue to set aside significant funds to fight pollution, notably air pollution, and that there will be coordinated efforts to reduce particulate matter concentration and treatment of ozone layer depletion.

Let me digress for a minute here. China is not the only country struggling to balance radical economic progress and environmental safety and conservation. Often, governments do not take sufficient measures until the situations become so terrible that not taking the steps would lead to complete political upheaval. Consider London, which was so enamored with economic progress that it allowed air pollution to continue unbridled up until the Great Smog of 1952. The smog was so stark that it forced policymakers to pull up their environmental socks and bring about the Clean Air Act of 1956—which also went through many amendments before it reached a publicly acceptable level. In America, the Environmental Protection Agency introduced the Clean Air Act in 1963 as a purely *political measure.* The Clean Air Act of 1963 was the very first federal law addressing air pollution management. It created a federal program under the Public Health Service and allowed research into monitoring and regulating air pollution. In 1967, the Air Quality Act was established to increase federal government operations. In line with this legislation, the government started enforcement in regions susceptible to statewide air pollution transport. As part of these procedures, the federal authorities undertook comprehensive ambient monitoring investigations and static source inspections for the first time.

So, while the intentions for introducing environmentally sound legislation are definitely up for debate, one can say without hesitation that they contribute to lowering pollution levels. The unfortunate thing is that in both the above cases, the reduced levels of pollution came at the cost of economic instability and the loss of livelihoods. For developed nations that are far better placed in the economic tapestry of the world, this may not be of significant consequence. However, for countries still in the race to establish themselves financially, the tradeoffs representing the fragile relationship between protecting the environment and allowing economic progress can lead to a significant hassle and, more alarmingly, public unrest. Economic development in such countries is often majorly dependent on the outputs of the manufacturing industry and the development of other areas like urban upliftment and locomotive trading. So, it was natural for China to want to focus on the economy first.

The Communist Party of China, which is in power, has as its overriding aim the establishment of a modern socialist nation to guarantee the well-being of its people. That is a philosophy that has necessitated continual economic development while also increasing environmental degradation. To break the connection between growth and emissions, policies that target fossil fuels while also encouraging the adoption of renewable energy sources will be required. If the shift results in inflationary pressure, it will be necessary to alter monetary policy accordingly.

What is important is that now, they are taking steps to lessen the damage inflicted on the environment. Even though China has the world's second-largest economy, it is still classed as a developing country and has not achieved its peak emissions levels yet. That is expected to happen by 2030, with Chinese President Xi Jinping pledging to achieve carbon neutrality by 2060. If China achieves this objective, it would have the quickest drop from maximum emissions amongst major economies, much faster than Europe's aim of seven decades and the United States' goal of four decades, respectively. As a result of China's proposal—which, according to the country's climate envoy, covers all greenhouse emissions—worldwide efforts to control global warming would be boosted, and the country could have more influence in international affairs.

China's government is focusing its efforts on reducing domestic solid fuel use in some of the country's most polluting regions. China must discover other energy sources to replace the carbon fuels that have driven its economic growth and fast urbanization. In July 2021, China launched the world's biggest carbon trading market, establishing a structure for how toxins are priced and controlled in the nation. This was an essential first step in the government's climate change efforts. It is already promoting electric cars and automation and expanding nuclear energy, which does not produce greenhouse emissions, as part of its climate change strategy. Research on innovations such as battery storage and hydrogen fuel to supplement low-emission energy sources is receiving increased funding. A proposal to substitute coal furnaces with natural electric and gas hydronic heating in roughly a million households across eleven cities was announced by the government in October 2018. However, in response to a shortage of natural gas, the Energy Development Administration changed course in July 2019 and suggested a thermal plan tailored to meet various regions' needs.

The Chinese government has taken tangible measures to reduce air pollution caused by coal combustion. The country is retiring many outdated, less economical coal power stations and substituting them with ultra-supercritical units intended to generate more electricity with less coal than previous plants.

All of its major institutions share China's aim of being carbon neutral, and it is a high priority for the People's Bank of China (PBOC). The central bank announced that it would eliminate clean-coal initiatives from its classifications of green bonds while also committing to updating instruments to provide banks with low-cost funding to encourage them to make more ecologically friendly loans. Regulators also want to change the regulations governing capital adequacy and the way green assets are counted. As of 2020, China's outstanding green loans totaled 14 trillion yuan, a figure expected to grow in the coming years.

In January 2017, China pledged to spend $367 billion on renewable power generation such as solar, wind, hydroelectric, and nuclear energy through 2020. Regulations in China are not just focusing on big enterprises, but they are also attempting to reduce negative environmental impact from other sources. One of the most important of these is automobile emissions. Beijing, Shanghai, Tianjin, and the Hebei region are among the highly populated areas in China where the sixth step of motor vehicle emissions regulations is in enforcement, commencing July 2019. The China VI regulations, as they are referred to, require cars to have improved exhaust gas filtration systems to capture more pollutants.

In metropolitan areas, better air quality will reduce premature mortality from fine particulate matter and ground-level ozone exposure. These regulations would reduce premature deaths and hospitalizations by at least twenty-nine thousand per year by 2030. Research estimates that the annual health benefits of implementing China VI will be 57 billion dollars (or 378 billion yuan) in the year 2030, with a gain ratio of 21:1 and net benefits of 54 billion dollars (or over 300 billion yuan)—a very cost-effective standard. The technology upgrade cost will be about three billion dollars (or a little over 18 billion yuan).

It is technically possible and ecologically beneficial to power ships moored in port facilities with electricity supplied by the grid rather than with their diesel engines, saving fuel costs and pollution. Not only would this increase the efficiency of their operation, but it also makes it cleaner and less damaging. This has been actively promoted in China via a combination of government incentives, statutory regulations, and demonstrations through pilot projects.

China has also contributed funds to the deployment of alternative-energy vehicles. The nation is projected to fade out fossil fuel cars by 2050 fully. The general and specialty vehicle sectors likely to accomplish this objective first, followed by the rest of the economy and the private sector. The State Grid Corporation of China (SGCC) is making significant contributions to the electrification of ports and harbors. By the end of 2018, the SGCC had completed the installation of electrical distribution systems in all public service locations along with the Beijing-Hangzhou Grand Canal system.

To make a significant impact on global oil consumption- an electrification effort must go beyond Tesla and municipal buses. According to Bloomberg, mainly in China, a cumulative 270,000 barrels a day of diesel demand will have been displaced by electric buses. That's more than three times the displacement by all the global electric passenger vehicles (Nightingale, 2019). This is mainly consistent with the belief that municipalities and even national governments have the authority to determine which buses may be purchased for use on public transportation networks in their respective cities.

In China, electric buses account for 98% of all buses. More than five hundred thousand have been deployed to Chinese streets as part of a concerted campaign through the Chinese government, whereas only a few are used in other parts of the world. This has a significant impact on oil displacement: China's electric buses ended up saving more than two hundred thousand barrels of oil per day in 2018, at a time when there were 400,000 electric buses on the road in China. By comparison, electric vehicles all over the other parts of the world saved only around forty thousand barrels per day in 2018 (Nightingale, 2019).

In recent years, electric cars have gained prominence among fleet managers, and Beijing has taken the initiative to go green with the advent of the electric taxi. The city is home to one of the most significant cab fleets globally, about seventy thousand vehicles strong. As part of a new air pollution control project initiated in 2017, the cabs have been converted to electric cars. The overall cost of the fleet conversion is anticipated to be about 9 billion yuan or 1.3 billion dollars, but it is well worth it in the long run. Electric vehicles have the potential to save taxi businesses money on petrol expenditures while also reducing maintenance costs. They will also aid in preventing pollution in the city since the majority of Beijing's population is exposed to polluted air regularly.

In 2020, China created a high-capacity electric locomotive, which was a significant step forward in the nation's locomotive technology. Not only faster and more powerful, but this locomotive is also more environmentally friendly, safer, and more intelligent than its predecessors. The locomotive, dubbed "Shen-24," can haul a 10,000-ton cargo train up a 1.2% grade at 120 kilometers per hour. The locomotive will be used on the Shenmu-Shuozhou Railway in the coal-rich regions of Shaanxi and Shanxi once it is deployed. This is one of the primary energy transportation routes.

Chinese railway electrification has progressed remarkably over the past two decades, with most of the country's railroads being electrified. Passenger and freight trains accounted for 75% of all rail traffic in 2016, with electric trains accounting for the remaining 25%. This is a greater than threefold increase over the previous year when the number was just 21%. Shenzhen, a technological center in southern China, has surpassed most other major Chinese cities regarding gas-powered taxi conversion. The city of Shenzhen will only allow completely electric cars to be licensed as online ride-hailing taxis. The city hopes to complete its shift to all-electric taxis within the next few years.

Baggage cars, ferry buses, and other Beijing Daxing Airport Terminal vehicles are powered entirely by electricity, as is the whole airport complex. Aside from that, when aircraft are linked to the jet bridges, they get their electricity from the public power grid. This, too, is a significant investment, but one that will produce environmentally sound returns.

China has also introduced the concept of Green Finance or Green Credit. To put it simply, green finance utilizes funds, private loans, and micro-lending to advance sustainability. Green finance aims to incentivize people to do ecologically beneficial activities by making it financially worth their while. There is something for everyone, from significant investments in climate-change mitigation technologies to small-scale home internet loans for customers available the following day. This enables everyone to contribute to the promotion of sustainability in their own way. The majority of green loans are linked to particular performance criteria.

Green Finance offers the financial instruments that active actors need to create activities that have positive and long-lasting externalities in a growing number of cases. Examples of Green Finance initiatives include: encouraging alternative fuel utilization, clean energy use, freshwater sanitation, ecological audits, the mitigation of industrial and transportation pollutants, negative climate change, deforestation, and reducing one's carbon footprint. Its key benefit is that it involves the private sector in financing environmental projects, particularly filling gaps left by inadequate public budgets.

By the close of 2020, China has already reached an outstanding green loans figure of 1.8 trillion dollars (or 12 trillion yuan), the highest that any country has invested. It is poised to become a global powerhouse in green finance policy internalization. Environment-friendly finance regulations play an important role in directing the provision of credit to businesses in under-developed financial ecosystems, and state-owned companies. Up to this point, green financing has benefited China's efforts to protect the environment. China's green financing initiatives have had a significant impact on the reduction of industrial gas emissions.

It has also been shown that green finance regulations have a beneficial effect on the manufacturing sector, as seen by the positive connection between green financing instruments and company innovation. Green credit policies have been shown to enhance economic performance while simultaneously reducing pollution emissions substantially. FinTech firms in China provide a support layer for the financial services sector by using emerging technologies such as artificial intelligence (AI), cryptocurrency, cloud services, data science (ABCD), and the Internet of Things (IoT). FinTech is set to play an essential role in the supply of green financing in the following decades, using predictive analytics and machine intelligence to promote a sustainable transition among households and small and medium-sized businesses. FinTech development helps protect the environment investment efforts and encourages the reduction of sulfur dioxide emissions, positive outcomes.

The government has also invested in several programs to allow for development and environmental conservation to go hand-in-hand. The Innovative Financing for Air Pollution Control in Jing-Jin-Ji Program was China's first operation to use the Program for Result (PforR) plan in the energy sector. It was the first operation in the world to do so. The World Bank created the Program-for-Results (PforR) funding mechanism to meet the increasing need for achieving long-term results and establishing institutions. Another distinctive aspect of the initiative is that its funding is tied to accomplishing specified project goals rather than leveraging external institutions or procedures. Encouraging collaborations and matching development partner objectives and outcomes with national policy promotes social welfare programs and help utilize World Bank development assistance.

In China, the goal of adopting the program was to assist in the implementation of the government's Air Pollution Control Action Plan as well as the thirteenth Five-Year Plan memorandum (2016-2020) for energy conservation and clean technology, with a particular emphasis on controlling air pollutants at their source through energy efficiency improvements and the expansion of renewable sources. As a result of the initiative, the government has shifted its emphasis to controlling harmful emissions at the point of generation by decreasing coal use and boosting the use of clean and sustainable energy sources. In addition, this has resulted in significant reductions in greenhouse gas emission emissions.

As part of the program's implementation, the World Bank has worked closely with China's National Development and Reform Commission (NDRC) and the National Energy Administration (NEA). The International Bank granted a five hundred million dollar loan for Reconstruction and Development (IBRD). The IBRD money has multiplied the initial loan amount by more than five, allowing the government's air pollution reduction project to receive much more cash. With an emphasis on enhancing the program's outcomes monitoring and verification system, increasing capacity, and creating new financial products, the Global Environmental Facility (GEF) has also awarded a grant to help execute it.

Through the mainstreaming of green finance in financial institutions, the initiative was able to bring in much-needed funding, allowing the government to meet its goals. In this case, the initiative assisted a financial institution, Huaxia Bank, in providing funding to businesses reducing air pollutants and carbon dioxide emissions by improving energy efficiency, investing in renewable energy, and reinforcing air pollution controls as part of their business operations. By creating a Sustainable Finance Center, implementing green lending processes, testing new funding models and commodities, and educating its employees, the initiative has helped Huaxia Bank build its institutional capability.

The Program for Results has, since its inception in 2016, invested in twenty-seven energy-efficient, sustainable sources and emission control subprojects. Over a billion dollars have been invested, with the World Bank providing $208 million, Huaxia Bank providing $430 million, and sub-borrowers providing the rest $662 million. They leveraged the borrowed money more than five times.

Huaxia Bank has made environmentally friendly lending a priority. The bank has created a green finance unit, and the firm's six main strategic business lines now include green financing. Green finance has risen to over a hundred billion yuan in the company's operations as of 2019. The program has also reduced carbon dioxide output in the Jing-Jin-Ji area by over two million tons per year since its inception, which helped the country's attempts to slow climate change. Additionally, seventy-three megawatts of novel distributed solar present value (PV) business models were funded. In addition, financing was provided for installing distributed solar present value (PV) systems in over two thousand rural homes as part of a pilot project.

Additional reductions in coal use have been made due to better use of renewable energy in mining and manufacturing and an increase in the amount of renewable energy available. As a result, pollution abatement procedures such as the setup of particle matter-eliminating technological equipment, desulfurization, nitrogen fixation of combustion products, and swapping coal for natural gas have all come into play. The new Green Finance Center is expected to have sponsored green projects worth about a hundred and fifty billion yuan (about twenty-one billion dollars) by the end of the program's execution in 2022.

CHAPTER SUMMARY

- The first fund in China exclusively devoted to ecological improvements collected up to 88 billion yuan during its first phase to increase investments in green ecologically sound initiatives and commercial companies.

- Between 2016 and 2020, the Chinese central government committed 97.4 billion yuan to combat air quality decline, 78.3 billion yuan to combat water contamination, and 28.5 billion yuan to enhance and relieve soil. Furthermore, it has spent 20.6 billion yuan to combat environmental problems in rural areas.
- China's outstanding environmental loans totaled 14 trillion yuan as of 2020—a number projected to rise in the future years.
- China committed $367 billion in renewable energy production such as solar, wind, hydropower, and nuclear energy until 2020 in January 2017.
- China will create the world's largest carbon trading market in July 2021, creating a framework for how pollutants are priced and regulated in the country.
- In 2018, China's electric buses saved over 200,000 barrels of oil per day, even though there were 400,000 electric buses on the road. In contrast, electric cars in the rest of the globe saved just around 40,000 barrels per day in 2018.
- China built a high-capacity electrical

locomotive in 2020, which was a significant step forward in the country's locomotive technology.

- The Innovative Financing for Air Pollution Control in Jing-Jin-Ji Program was the first operation in China and globally to utilize the Program for Result (PforR) plan in the energy sector.
- Since its start in 2016, the Program for Results has invested in twenty-seven energy-efficient renewable sources and pollution control subprojects. Over a billion dollars have been committed, with the World Bank contributing $208 million, Huaxia Bank contributing $430 million, and sub-borrowers contributing the remaining $662 million.

In the next chapter, you will learn how legislative policy and implementation have helped China move forward in achieving its environmental goals.

6

TIGHTER REGULATIONS

While the availability of comprehensive, transparent data and increased funds have helped a lot in environmental policy formation and funding, China has not stopped at just this. It has also imposed stringent regulations, rules, and memorandums on large-scale corporations and small-scale enterprises.

Combined with the fast development of renewable energy generation, China's increasing openness to multilateral methods to greenhouse gas emission reduction has earned worldwide acclaim and appreciation for the green entrepreneurial state of the country. In the wake of the Kyoto Protocol agreement, China was severely condemned by players in the international political arena for what they believed to be its stubbornness in climate change talks. There was a time when it was seen as a reckless polluter committed to economic development at any cost inside its borders. Fast forward to now—China had overtaken the United States as the world's largest generator of renewable wind power in 2010 and as the world's leading installer of photovoltaic cells since 2013.

Ever since there has been a shift in the economic model to stop and look at the importance of the environment, China has come up with several sound initiatives. Significant among these is the *Two Mountain Road,* which believes that one can still go after economic development while looking after the environment and everything it does to sustain humankind. Individuals who damage the environment excessively and incur substantial economic and health expenses, as a result, are the primary targets of fines. Additionally, in more severe cases, a court will investigate the legal responsibilities of the perpetrator to discipline the offender.

China is now generally recognized as the global leader in renewable energy. Aside from that, Chinese factories produced more than six hundred and eighty thousand electric vehicles (cars, buses, and trucks) in 2017, exceeding the entire global output of the same vehicles. Under President Donald Trump, China has developed a new green image that contrasts sharply with the United

States' departure from the worldwide environmental change frameworks and disdain for renewable energy development, among other things. The Chinese domestic decision to work on environmental conservation has recently resulted in substantial improvements in air quality in certain parts of the country, notably Beijing, as reported by global news media and ecological non-governmental organizations (NGOs).

China's environmental law has been strengthened in specific sectors, including air quality, water or soil degradation, and solid waste disposal. Regional limitations on total contaminant discharge amounts have been established throughout the country in various areas. The government has started accumulating environmental damage fees commencing January 2018. The cost is determined by the number of pollutants emitted. The former Ministry for Environmental Protection has now become the Ministry of Ecology and Environment. It is now a broader and more powerful entity with overall responsibilities and is much more capable of enforcing environmental rules and regulations. It is responsible for all policies surrounding environmental issues and climate change.

China has added a significant number of wind and solar power plants to its electrical system, and major domestic businesses for the production of solar panels, batteries, and electric cars. To close out 2017, the country implemented a nationwide emissions trading scheme, which establishes a market for purchasing and selling carbon dioxide emissions permits. Considering the fact that the United States has yet to develop a national market-based climate policy, this was a highly significant step.

Environmental bureaucratic reforms adopted in 2018 are intended to reduce inefficiencies and conflicts that occur due to the substantial dispersion of government policy and enforcement authority in the environmental sector. When the Ministry of Environmental Protection was mothballed as part of the government's dynamic restructuring of critical ministries in 2018, the Ministry of Ecology and Environment was created to take its place.

It is hoped that the recent introduction of an environmental tax in line with a revised Environmental Protection Tax Law would result in a significant rise in the expenses of damaging the environment. Under the reform, environmental responsibilities have been assigned to local tax authorities, who have the power to collect an environmental protection tax in case of damage or deviance from conservation rules. This would take over the role of the existing charge system. It is widely expected that this move will force local governments to stop exploiting loopholes and protecting harmful companies that provide the most significant amount of political tax revenue and employment outcomes. The new environmental tax revenue can be included in local fiscal profitability, giving chronically income-deficient local governments an incentive to improve the methods for collecting environmental taxes as part of a holistic fiscal strategy.

Twelve provinces and municipalities have put environmental protection policies in place since September 1979, including Shanxi and Shandong. These policies are in accordance with relevant sections of environmental protection law. Fines have been imposed in the provinces of Hebei and Liaoning for polluting the environment. In some regions of the Jiangsu, Anhui, and Shanxi provinces, the use of fines has been tested, and the results are encouraging. Up to this point, the regulations have been effectively applied in more than one thousand five hundred companies and government entities in over fifty cities, eleven prefectures, and fifteen provinces, with favorable outcomes. Overall, penalties for violating the state-imposed limit for the emission of pollutants are marginally more costly than the costs of pollution prevention and control measures themselves. As a result, companies are shown that investing in environmental management is better than paying fines. The motivation to start supervision as soon as feasible rather than later also helps provide the conditions required for maintaining the production process up to date and reducing waste throughout production.

Following the implementation of the new procedure, both cadres and regular citizens have become more familiar with environmental safety regulations. They have started to comprehend that they are financially liable for any environmental damage caused and must not take the matter lightly. The fines will also contribute to the supply of funds to advance environmental management, which is a positive development.

The establishment of ecological police (EP) squads in many locations has been a significant development. Although they established administrative procedures and regulations, many did not get the attention they deserved from the authorities overseeing their implementation and enforcement. Some policies and rules were overlooked or ignored altogether. The trend of enforcing fines for polluting emissions has put an end to this indifference and aided in developing technological components and surveillance skills in the environmental sector.

Since establishing the country's first EP force in Kunming, China's local and provincial governments, including those in Guangdong and Beijing, and also

Hebei, Sichuan and Shandong, Anhui and Jiangsu, Yunnan, Shanxi, and Hubei have all joined the bandwagon. China's environmental regulators believe that by pursuing environmental degradation cases as criminal offenses, the manufacturing industry would be more driven in the future to comply with the country's environmental laws and regulations. For instance, the obligatory procedures undertaken by the policemen as part of the EP squads may entail jail, which would deter polluters from continuing their activities.

The Ministry of Ecology and Environment, as already mentioned, is more prominent and potentially much more powerful than it had been previously. The new MEE has been given increased authority for policies associated with climate change that were previously within the competence of the National Economic and Development Authority. This is in addition to its duties for pollution control. In addition to helping to address China's long-standing problem of disparate conservatism, this realignment of administrative authority and responsibility at the government level may also help to address enduring issues stemming from the blurry coordination of activities between departments. This realignment will integrate power under the MEE's guidance and resolve recurring duplication and conflicts in political decision-making and mitigation strategies.

A total of eleven regulations and memorandums targeted to local governments and industrial sectors have been published by different government agencies since January 2017, requiring energy consumption reductions, carbon emissions reductions, and tighter environmental restrictions. To enforce implementation, the national government has undertaken three rounds of thorough environmental inspections. The 2017 environment protection act in Beijing, Tianjin, Hebei, and adjacent areas is the most significant of these regulations, with the most excellent effect on the energy and industrial industries. China's Ministry of Public Security and the Municipal Government of Beijing jointly published the policy, dubbed the "2+26 Act," which focuses on Beijing, Tianjin, eight cities in Hebei, four cities in Shanxi provinces, seven in Shandong, and seven in Henan province.

The views of individuals who have maintained a lengthy role in China's governance systems regarding the environment have changed due to the inflow of new players. For example, higher-ranking Party cadres in critical administrative units that cross inter-jurisdictional river basin boundaries are being assigned objectives for quality of water to improve the overall management of water throughout the country. The regional or municipal governments appoint river chiefs. River chiefs are now higher-ranking cadres at the prefectural or municipal level who are tasked with overseeing lower-ranking cadres in charge of tributaries that circulate into the river basin. Due to the transfer of responsibilities from the government to party organizations, party leaders are gradually taking on important environmental management duties.

To combat the country's deteriorating air quality, China started strengthening environmental restrictions in January 2015, revising the country's environmental protection legislation for the first time in nearly three decades. To provide the government a greater role in environmental protection, leaders crafted the new legislation with resolution. Examples include the need for the government to develop obligatory emission regulations in the future. In addition, the new legislation eliminates a limit on the amount of money that may be charged to errant organizations. In addition, the legislation has a system that allows conservation groups and the like to sue corporations for liabilities that may result from any form of environmental contamination.

Since it is doing so much to protect and enhance the national environment, it is natural that China has now positioned itself as a power to reckon with in climate action. It has made investments that have brought renewable and alternative energy sources into the limelight and mandated economic and legislative policies to achieve sustainable development. Reports at the newspaper *The Conversation* based in Australia note that over the last decade, China has implemented over a hundred policies that have been instrumental in reducing the consumption of fossil fuels and lowering greenhouse gas emissions (Gallagher & Zhang, 2019). These policies will have mostly beneficial side effects, such as improving China's energy security, stimulating economic growth, and reducing ground-level pollution, to name a few. Only the emissions trading system is explicitly focused on reducing carbon dioxide emissions, making it the most critical effort in this area.

Of these hundred, a good instance is a feed-in-tariff policy aimed at renewable energy generators. It is common for Feed-In-Tariff schemes to ensure that customers who own renewable energy generating facilities eligible for FITs get a fixed fee. Per the FIT scheme, customers with roof-top solar photovoltaic systems would get a fixed fee from their utility for all the power they produce and provide to the grid. Onshore wind, solar PV plants, distributed solar, and offshore wind sectors have all benefited from China's feed-in tariff (FIT) pricing system, which has been in place since 2009, 2011, 2013, and 2014, respectively. Within this FIT system, the national pricing regulator, the National Development and Reform Commission (NDRC), established various FIT rates for different renewable projects, which reflected the varying costs associated with tapping into these resources. It is a cost-based model that considers the average residual project cost and a set internal rate of return (IRR), typically 8% in most cases. With feed-in tariffs, the price of renewable energy producers' electricity is now guaranteed. Energy efficiency requirements are now in place for power plants, cars, buildings, and equipment; non-fossil energy production goals are set, and limits on the consumption of coal are required.

In March 2017, the Department of Environmental Protection and the National Development and Reform Commission (NDRC) was introduced. They want to reduce energy consumption, enforce stricter waste discharge permits, implement more stringent regulations on volatile organic compounds (VOC) emissions, and eliminate redundant capacity (supply-side correction). They also want to conduct thorough environmental impact and energy-saving assessments before approving any large project.

This has the impact of removing excess supply chain capacity. The main goal of pollution control is to reduce capacity, but the unintended consequence is a significant influence on the renewable energy market—which, in turn, will help improve its profitability.

Often, the reason why pricing rivalry favors conventional energy sources (like coal mining, steel production, and glass production)—and polluting sources of building materials *and* the use of environmentally harmful chemicals—is that they are readily available in bulk. This makes them far cheaper compared to renewable sources, which are not available as quickly. Changing this will turn some of the competition in the latter's favor.

The implementation of legislation under the NDRC produced a significant effect on the energy industry. Coal has long been the most widely used fuel for electricity generation and heating. The use of coal is encouraged by the federal government as a method of combating air pollution. It is mandated that coal be replaced by natural gas or electricity in areas with high population densities. This policy has an immediate impact on the energy industry—the requirement for natural gas increases. Natural gas consumption is anticipated to rise even higher in North China during the colder months. Some major energy firms are investigating ways to increase gas supply from all sources, including increased domestic production, gas imports through pipelines, and coal-bed methane yield.

In 2020, China enacted the Energy Law to make energy policy more controllable and reduce commercial energy use yearly. The goal for 2020 is to cut emissions by 2.5% by the end of the year. To compare energy consumed via electricity, China used 6840 terawatt-hours of power in 2018, which rose to 7225 in 2019. In 2020, the rise was to 7510. Between 2018-19 was 385 TW, while the increase between 2019-2020 was 285 TW (*Electricity Consumption in China between 2010 and 2020(in Terawatt Hours), 2021*).

In terms of the environment and economics, the ideal approach is limiting emissions and charging for them. With cap and trade, one may achieve both objectives at once. It's a mechanism designed to help the environment by cutting down on pollutants. A government regulatory system known as cap and trade aims to limit or cap the total quantity of particular chemical contaminants, such as carbon dioxide, caused by industrial activity. The government gives out a certain number of licenses each year to companies to emit a specific amount of carbon dioxide. As a consequence, the maximum amount of emissions is equal to the entire amount permitted. There will be a financial penalty for businesses that release more carbon dioxide than their permits. It is possible for companies that reduce their CO2 emissions to trade and sell their extra permits with other companies that want to buy them.

Strict limits on greenhouse gas emissions, which contribute to global warming, are used to safeguard the environment. The upper limit is tightened with time. A market for trading permits allows companies to emit a limited amount of carbon dioxide, with the price set by supply and demand. This is the trading part. Companies that participate in trade have a strong incentive to reduce pollution in the most cost-effective way possible.

With this in mind, China has also introduced a country-wide carbon trading industry known as the National Emissions Trading Scheme, which came into existence in 2021. It initially evaluated and monitored carbon emissions arising from the power sector, but with time, it is also expected to expand supervision over the construction industry, chemicals and oils, the aviation industry, and iron and steel industries. The primary aim is to reduce the intensity of pollutants being emitted rather than doing away with them altogether—which is practically a risky thing to take up when so much of the economy still needs to be developed. This scheme will also study and evaluate target reductions per unit output, weighted heavier than total carbon dioxide emission. It gives them the incentive to produce more environmentally efficient and clean energy that is not dependent on the intense burning of carbon fuels, which will be the road to further reducing pollution. China may welcome foreign players from the United States and Europe to use this platform to trade carbon reduction credits.

CHAPTER SUMMARY

- Recently proposed reforms to the environmental bureaucracy in 2018 aim to eliminate waste and conflict caused by the uneven distribution of government policy and enforcement authorities in the environmental sector. They are expected to take effect in 2019.
- When it comes to renewable wind power generation, China surpassed the United States in 2010, and when it comes to solar cell installation, China has eclipsed the United States since 2013.
- In the field of renewable energy, China is now widely regarded as the world's leader. Aside from that, Chinese companies produced more than 680,000 electric vehicles in 2017, outnumbering the total global output of the same vehicles.
- Establishing ecological police (EP) units across the country has been a significant development in the last few decades.
- In comparison to its prior size and potential,

the Ministry of Ecology and Environment has grown significantly. The new MEE has been given expanded control over climate change policies that were formerly the responsibility of the National Economic and Development Authority.

- Due to China's poor air quality, environmental regulations were stepped up in January 2015, becoming the first in nearly three decades that China had updated legislation governing environmental protection.
- Since January 2017, several government agencies have produced eleven ordinances and memorandums aimed at local governments and manufacturing industries, necessitating energy consumption cutbacks, carbon emission reductions, and harsher environmental controls.
- China's energy law will go into effect in 2020, requiring companies to lower commercial energy consumption annually.
- China likewise established in 2021 a national carbon trading business known as the National Emissions Trading Scheme. It first examined and monitored carbon emissions from the electricity sector, but it is anticipated to expand oversight to include the building industry,

chemicals and oils, the aviation industry, and the iron and steel industries in the future.

In the next chapter, you will learn all about China's present-day status in terms of the environment and how far they have mitigated emissions and reduced air pollution.

7

CHINA'S ENVIRONMENTAL STATUS TODAY

When it comes to air quality, it is impossible to escape pollution no matter how prosperous an area you live in. You may find it in abundance all around us. Contaminants in the air that penetrate deep into our mitochondrial respiratory chain and cause damage to organs such as the lungs, heart, and brain may be able to overcome our systems' natural defenses against microorganisms in the air, causing illness.

Carbon dioxide emissions from fossil fuel combustion contribute to climate change and air pollution, and an attempt to decrease one may positively impact the other. Climate change and air pollution are inextricably intertwined. According to the United Nations' Intergovernmental Panel on Climate Change, we must phase out coal-fired electricity by 2050 to prevent global warming from becoming catastrophic. Without a significant shift in behavior, we may be confronted with a devastating climate disaster sometime in the next two decades. The problem of air pollution affects the whole globe. It was responsible for more than a hundred thousand untimely deaths in the United States in 2011 and cost the nation approximately nine hundred billion dollars. Carbon dioxide pollution is intimately related to climate change; fossil fuel burning is the primary cause of climate change and a significant contributor to air pollution; therefore, reducing one may help lessen the other.

Community backlash has been the motivating factor behind China's pollution policy, and a wave of demonstrations against the country's pollution policies has arisen in recent years. The government is forced to perform a delicate balance while dealing with an economic system that has already accelerated efficiently and rapidly without the emissions standards and protocols that people now demand. It must address the broader public environmental concerns while avoiding actions that would slow growth.

According to current estimates, more than five million people die pre-emptively each year due to cardiovascular diseases such as cancer and heart disease, and diabetes, and breathing problems caused by pollution in the environment. Compared to pollution-related deaths, Dengue fever, tuberculosis, human immunodeficiency virus infection, and acquired immune deficiency syndrome cause fewer deaths, even when combined. Air pollution exposure may harm the health of children, the elderly, people who are already ill, and minorities and low-income collectives, who are particularly susceptible to serious treatment outcomes and economic consequences, such as lost workdays, due to the pollution.

Of course, the truth is that China has prioritized economic development for far too long. One cannot blame this desire solely because it is a singular desire of all developing nations to reach a stage where they are financially stable and have a proper hold on the power play of international politics.

By regulating emissions in individual provinces, China began to take pollution management seriously around 2006. The first effort resulted in an overall reduction of more than 10% in Sulfur dioxide emissions throughout the country compared to the year 2005. A study of compliance over time also suggests that air pollution reductions began to take place with rigor starting from 2008, shortly after China's government established the Ministry of Ecology and the Environment (MEE), which used to be identified as the Ministry of Environmental Protection. Many environmental policy changes occurred after the MEE's establishment, including collecting accurate sulfur dioxide atmospheric emissions data from continuous emission surveillance systems (CEMS) at the county and state level and an increase of 17% in terms of numbers of enforcement officers.

To address air pollution issues, the Chinese government has developed several policies designed to improve air quality. Following 2005, these policies have become more and more comprehensive. In its eleventh five-year plan, China established a target of reducing nationwide sulfur di-oxide emissions by 10% in 2010. In the next five-year plan memorandum, the goal was increased by 8%, totaling a reduction of 18% in total sulfur dioxide emissions and a further 10% reduction in nitrogen oxide emissions by 2015. In 2013, China's State Council came up with the Air Pollution Prevention and Control Action plan, one of its strictest plans to date. The program mandated reductions in particulate matter concentrations between 15-25% in all key regions, including Beijing, by 2017.

By all counts, China is doing a better job of cleaning up its air than Britain did after the Industrial Revolution. Even while air quality in England is far from ideal, it had improved significantly since the days of the Great

Smog, when thousands perished due to poor air quality. According to the World Health Organization, as early as the 1980s, air pollution levels in major Chinese cities were almost identical to those experienced by London at the zenith of the Industrial Revolution in 1890. In contrast, poor air quality in China was declining at a rate comparable to that of London ninety years ago, but twice as rapidly. As is usual for an increasingly industrialized economy, China's recent past has seen high levels of air pollution; nevertheless, the country's progress in flushing air pollution has been rapid relative to historical norms.

From 1998 until now, China has substantially expanded its economic development in rural development and environmental protection. In sixteen sustainability projects spanning upwards of six hundred million hectares of land, representing above 60% of the country's total land area, more than three hundred billion was invested through 2015. It is an endeavor that, although not without faults, is unrivaled elsewhere in the world. To minimize environmental impact, some of the environmental objectives were preventing soil erosion and flash floods in the Yangtze and Yellow rivers. Other goals included Preserving forests in the northeast, significantly reducing drought conditions in the dry north and rocky south, reducing the impact of sand storms on Beijing, and ramping up agricultural output in China's central and eastern regions.

It was as crucial to achieving the political and social objectives of poverty alleviation and economic development, particularly in western China, as it was to achieve military objectives. As a consequence of schemes that paid farmers for implementing sustainability initiatives on their land, farmers' livelihoods improved. Housing and off-farm work in China's expanding metropolis not only helped to raise family incomes but also served to alleviate pressure on the government's limited land resources. China's developing cities are also a source of inspiration for entrepreneurs across the world.

We have already looked at China's work to fight negative climate change in the last few chapters. I'm going to take a minute to go over the crucial bits. Sustainability became a priority for the Chinese government in the late 1990s due to the dire circumstances that prevailed among the country's rural residents and in their immediate environments. Even though China had been planting crops for more than eight thousand years, the repercussions of unproductive and unsustainable agronomic systems and overexploitation of natural resources led to massive poverty and environmental debasement by the mid-19th century. Floods, droughts, and other natural catastrophes followed, a notable one being the Great Chinese Famine, which happened during 1959 and 1961 and culminated in somewhere between twenty million and forty-five million deaths, depending on estimates, as a consequence of a lack of food.

After the financial reforms of 1978, China established approximately six sustainability projects; however, with just a tiny amount of investment, the environment continued to deteriorate and eventually collapse. As early as the 1990s, natural forest cover had decreased to less than 10%. Approximately four billion tons of soil were also being lost per year, resulting in significant water quality and sedimentation problems in the environment.

Erosive soil loss exceeded hundred tons per acre per year in some regions of the Loess Plateau. According to the United Nations Environment Program, the Yellow River that flowed through the area had the unpleasant distinction of being known as the world's muddiest river. Over a quarter of China's agricultural soils had been destroyed by overgrazing, and the country's agricultural output had been falling for many years.

Unsustainable land management re-emerged in China in the late 1990s due to a series of natural disasters that the public generally acknowledged caused by improper land planning. Such as the Yellow River drought of 1997, the Yangtze River floods of 1998, and the catastrophic dust storms that plagued Beijing regularly from 2000 to 2002. Because of this issue of long-term viability, financing for environmental initiatives increased significantly after 1998, culminating in the launch of eleven new projects. Among the initiatives included in the portfolio was the *Grain for Green Program* and the *Three North Shelterbelt Program*, which aimed to stop and reverse desertification by building a four thousand five hundred kilometer-long Great Green Wall across northern China.

In 2019, renewable energy sources constituted about 15% of the sources it uses for day-to-day activities. This was an increase of 8% from what it had been almost ten years earlier. Over the years, China has consistently relied on hydropower as a good energy source for running industries and generating electricity. It has also worked on installing increasing numbers of solar power heaters and wind turbines and developing increased clean nuclear energy. It has become a global powerhouse of electric vehicles, ranging from cars to buses and even trains.

China is the world's leading producer of solar energy, outpacing every other country on the planet. In light of China's enormous population, that might not appear as significant in terms of quantity, but it is illustrative of China's future course. Wind power installations in China will surpass those in any other country by 2020, with a threefold increase over the previous year's total. Achieving a 25% share of non-fossil fuel energy production by 2030 is a goal China has set for itself, and several experts expect it will accomplish relatively soon.

Tsinghua University researchers estimate that usage of renewable energy will account for 90% of all power use in the country by 2050 (Myllyvirta, 2020). It is possible that China's technological leadership in renewable initiatives, such as photovoltaic arrays and large-scale batteries, will confirm to be a major asset in the pursuit of this goal. When green technologies were first introduced in China, they were intended to combat carbon emissions, which is a major problem in many cities. The aim now is to continue in this direction till the citizens of China have begun looking at alternative fuels as the primary sources of their day-to-day energy needs.

In 2020, over one million new energy-driven cars, including battery-operated electric cars, hybrid plug-in cars, and vehicles fueled by hydrogen fuel cells, will all become operative in China, totaling an increase of 11% from the previous year. Currently, China is ranked eighth in the world in terms of the proportion of electric vehicles sold in the country. Given its immense size, it manufactures and purchases many more electric vehicles than any other nation by a significant margin. By 2025, it also intends to manufacture batteries with double the capacity of those now produced by other nations.

Chinese officials say they are on track to meet their commitments under the *Paris Climate Accord*. It has been possible to reduce carbon intensity, usually expressed in carbon emitted per unit produced, by more than 40% since 2005. This has allowed the system to achieve a major goal two years earlier than initially planned. As a result of increased environmental regulations and massive green investments in the transition to electric vehicle infrastructure, the rate of increase in emissions has been significantly slowed in recent years. As a direct consequence, green marketing is gaining traction in the marketplace.

It has risen to become the world's leading producer of solar and wind energy capacity. It is the industry leader in the production of solar collectors and large-scale batteries. Aside from that, it produces more solar energy than any other nation.

China has the seventh-highest proportion of electric vehicle sales globally but considering its enormous size, China produces and purchases much more electric automobiles than any other nation. In China, about one out of every twenty cars purchased is an electric vehicle. Compared to other countries, China is becoming greener at a quicker pace. This is mainly due to the country's forestry projects intended to decrease soil degradation, flooding, and erosion.

In 2020, China's wind power installations will be more than three times greater than any other nation. In addition, China is home to more than one-third of the world's wind turbine manufacturing firms, more than 70% of the world's solar panels, and over 75% of rechargeable batteries, all of which are used to produce electric vehicles. China is likely to achieve its 2030 goal of obtaining 25% of its energy from non-fossil sources.

For more than two decades, China's efforts at preservation have been predominantly positive for the people involved. There has been a reduction in deforestation, and forest cover currently accounts for more than 20% of the total area. Recent years have seen a significant increase in the growth and regeneration of grasslands. Although climate change has significantly influenced shifting desertification patterns in many areas, restoration efforts have also impacted these patterns. Soil erosion has been substantially reduced, and both water quality and the deposition of excess sediment in rivers have improved dramatically as a result. The Yellow and

Yangtze rivers have both seen a 90% reduction in sediment loads. The increase in agricultural output has occurred as a consequence of increased efficiency and technological advances. Food insecurity has almost wholly disappeared from rural areas, and rural households tend to be better off on average.

Having said that, China continues to face many challenges. The levels of air pollution in China continue to be high, indicating that more measures are required. The particulate matter concentration is still higher than the world health organization's threshold for being deemed safe. Many areas continue to be plagued by long periods of severe smog, and hundreds of cities, especially in the north, continue to be plagued by high levels of particulate matter.

There is still a significant dependence on coal-fired electricity. In 2020, China will have added more new coal-fired power capacity than the rest of the world combined, according to the International Energy Agency. Despite this, it has failed to exploit its renewable energy potential fully. According to the National Renewable Energy Laboratory, in the three provinces with the greatest conditions for wind energy production, as much as 43% of the available wind turbine capacity is not being utilized.

Aside from that, China is the leading funder of fossil-fuel infrastructure and the world's largest coal exporter. Through its Belt and Road Initiative (BRI), it has committed to financing the building of dozens of coal-fired power plants. Nonrenewable resources have received more than 60% of BRI-specific energy funding. For the time being, Beijing's efforts to make the Belt and Road Initiative (BRI) more ecologically friendly while laudable is entirely voluntary. In addition, there is a need to keep up with urbanization and energy demand growth. With China's increasing urbanization comes greater dependence on high-energy-consuming goods such as cement and steel and increased use of automobiles. People in China possessed about 240 million cars in 2018, growing to about 27 million in 2004.

Even though it is still the world's leading emitter of carbon dioxide, its per capita emissions are less than half of those of developed countries. So, I will conclude this chapter by saying that while China has a long way to go, so do all the other countries facing the conundrum of reducing environmental pollution while also developing economically. What we should pay attention to is how much China has already achieved and how much *it is geared to do for the environment in the near future.* It has set ambitious goals, but given the pace and determination with which it is moving, it is hopeful that it will set an example in terms of cleaner air and environment—not just for its own people and developing countries, but also developed nations all over the world.

CHAPTER SUMMARY

- The issue of air pollution affects the whole world, and it was responsible for over a hundred thousand premature deaths in the United States in 2011 and cost the country about nine billion dollars.
- The motivation for China's pollution policy has been a community reaction, and a wave of protests against the country's pollution regulations has erupted in recent years.
- Around 2006, China started to take pollution control seriously by regulating emissions in specific regions. The initial attempt resulted in an overall decrease of more than 10% in sulfur dioxide emissions throughout the nation compared to 2005.
- To address air pollution problems, the Chinese government has enacted various regulations aimed at improving air quality, and these policies have grown more extensive since 2005.
- In its eleventh five-year plan, China set a goal

of reducing national sulfur dioxide emissions by 10% in 2010. The target was raised by 8% in the following five-year plan memorandum, totaling a reduction of 18% in total sulfur dioxide emissions and a further 10% decrease in nitrogen oxide emissions by 2015.

- By all accounts, China is doing a better job of cleaning up its air than Britain did during the Industrial Revolution.
- Renewable energy sources accounted for about 15% of the sources used for day-to-day operations in 2019. This was an 8% rise from approximately ten years ago.
- China is the world's top solar energy generator, surpassing every other nation on the globe. In light of China's enormous population, that may not appear to be a significant number, but it is indicative of China's future course.
- China's wind power installations will outnumber any other nation by 2020, with a threefold rise over the previous year's total.
- Over one million new energy-driven automobiles, including battery-powered electric vehicles, hybrid plug-in vehicles, and vehicles powered by hydrogen fuel-cells, would enter service in China in 2020, representing an increase of 11% over the previous year.

- China has the world's seventh-highest percentage of electric car sales, but given its vast size, China manufactures and acquires much more electric vehicles than any other country.
- China is on track to meet its 2030 target of getting 25% of its energy from non-fossil sources.
- For more than two decades, China's preservation initiatives have been overwhelmingly beneficial to the individuals engaged.
- There has been a decrease in deforestation, and forest cover now accounts for more than 20% of total land.

In the next chapter, you will learn how China has overtaken the United States regarding environmental conservation.

8

OVERTAKING THE US

Since 2012, China has made over three trillion yuan in investments to combat air pollution. Through legislative changes, the national government has expressed its commitment to environmental protection and welfare. A total of seventy articles have been added to the existing Environmental Protection Law, which went into effect in 2015. It gives precedence to the protection of the environment over advancement and grants the dissemination of information on environmental protection. Fines for illegal activities will be imposed under the new legislation, with no upper limit on the fine amount. In addition to environmental protection laws, the Water Pollution Prevention and Control Law regulates the treatment and release of water in urban, agricultural, and rural areas.

The state department also released a Plan Of Action for Prevention and Control of Water Pollution in 2015, intending to improve water quality in major river basins and urban areas throughout the state. The plan has ensured widespread coverage of the sewage network across countries and a minimum treatment rate of 95% in urban areas and 85% in rural areas, respectively. The action plan also aims to enhance facilities, recycle and regenerate industrial water, stop huge industrial spillages, and prevent pollution from livestock farming, among other objectives. A new Action Plan on Winning the Battle for the Blue Sky, published in June 2018, set out goals for joint administration of greenhouse gas (GHG) emissions while promoting air pollution control to achieve the blue sky. In the following year, the Ministry of Environment and Energy (MEE) issued the Inclusive Management Plan for Volatile Organic Compounds (VOCs) in Key Industries, which brought attention to the issue of VOC emissions. China has grown into a significant demand for industrial solutions for reducing VOC emissions from various sources.

Much of China's polluting production has its origins in other countries, thanks to the country's cheaper labor, abundant land, and lax regulations, which have attracted a large amount of foreign investment. The coal-fired power stations that feed China's economic engine, the businesses and infrastructure investments that keep the populace employed, and the Toyota and Buicks cherished by the flourishing middle class all contribute to China's pollution problem in significant ways. Chinese greenhouse gas emissions have surpassed those of the United States since 2007, accounting for 27% of global greenhouse gas emissions since 2007.

As part of its effort to demonstrate the superiority of its socialist economy, China has also amassed large-scale government assets that dwarfed those made by the United States, coupled with national mandates requiring infrastructure to use renewable energy. It put in place an end-to-end distribution network. The country now produces the vast majority of the world's polysilicon, a key component of solar panels. It has also managed to keep its labor costs down from most industrialized nations, and it has been prepared to support unprofitable businesses. Solar panels are currently supplied by Chinese companies in three-quarters of the world's total.

When it came to developing a domestic solar sector, China was using every resource at its disposal. Local governments supplied low-cost land, while state-owned banks provided favorable financing conditions. Beijing also stimulated demand for the goods by providing substantial subsidies, which helped it become the world's biggest consumer of solar panels and other renewable energy sources.

Chinese manufacturers were likewise attempting to increase productivity while decreasing costs. They made more solar cells from the same quantity of raw material thanks to new equipment that cut polysilicon wafers more thinly while generating less waste. Because of this invention, solar energy has reduced prices by 80%, making it as affordable as coal in many areas of the globe.

The influx of low-cost Chinese panels delivered a severe blow to American manufacturers, and Solyndra (a US manufacturer) wasn't the only one to suffer. After multiple other Solar photovoltaic manufacturers filed for bankruptcy protection, President Barack Obama slapped import duties on solar panels as high as over 200% in 2012. Large manufacturers circumvented the United States tariffs by constructing assembly facilities for solar cells and modules throughout Southeast Asia.

Compared to two decades ago, when 22% of solar energy cells were produced in the United States, just 1% is made on American territory now. At one time, there were seventy-five large solar components manufacturers in the United States, a figure that was anticipated to rise as the business grew. The majority of them have subsequently been closed. Only thirteen years after opening, the Hillsboro factory followed suit and closed its doors. Notwithstanding vast sums in government subsidies and almost two decades of promises from presidents, beginning with George Bush, that the United States would become a clean-energy behemoth, the sector has failed to establish a foothold in the United States.

The United States has emitted more greenhouse gases into the atmosphere since 1965 than any other nation. According to the International Energy Agency, the 278 billion megatons of carbon dioxide that the United States has released into the atmosphere during the last five decades alone accounts for about 21% of total world emissions during that period. By comparison, China's total carbon dioxide has slowed down (from 18% in 2011 to a little below 3% in 2014). China has made significant efforts to reduce the coal consumption of its economy, which is assessed in terms of the quantity of carbon released per ten thousand yuan of economic output in the country. During the years 2000-2014, a rise in Carbon dioxide emission was accompanied by a reduction in the country's carbon intensity of 57%.

Carbon dioxide emissions in the United States reached a high in 2007, standing at over seven billion tons. Carbon consumption intensity from coal combustion decreased as a proportion of total carbon dioxide emissions (from 36% in 1990 to about 32% in 2014), but this is mainly due to the significant rise in natural gas output, which has coincided with a drop in natural gas prices in recent years. Contrary to expectations, the proportion of Carbon dioxide emissions from natural gas usage rose by almost the same amount that it decreased by when coal was used. Per-capita carbon dioxide emissions in the United States are almost twice as high as those in China—seventeen tons of carbon per-capita placed against just above six tons in China in 2012.

When it comes to developing, monitoring, and enforcing strategies to meet national air pollution regulations, the United States has a lot of constitutional issues, making it more difficult to put its federal capability to use to achieve the Obama Administration's carbon reduction goal. In recent months, there has been significant resistance from certain states and some stakeholders who argue that the federal government doesn't have the power to unilaterally impose state plans to decrease carbon emissions under the Clean Power Plan. The Supreme Court of the United States just ruled against the Obama administration's stance, casting considerable doubt on the survival of the Clean Power Plan in the process. Compared to the United States, China has a better capability of requiring all branches of government to execute central government requirements, such as converting the nationwide carbon intensity goal into provincial-level objectives to aid in compliance with enforcement.

Here is the comparative analysis: in the United States, carbon emissions have been on the rise following the industrial revolution. In contrast, for China, there was no significant rise for most of the 20th century. However, from 1979, economic reforms in China began to take precedence over state impositions, and added to this, cheap labor also became widely available. China became the hub for most nations to carry out their offshore business production requirements at minimal cost while inflicting maximum harm upon the environment and the Chinese populace. Additionally, the 1979 reforms also loosened the limits on private industry and opened up Chinese trading systems to the rest of the world.

For obvious reasons, the combination of all of these led to a manufacturing boom and economic prosperity. However, it came at a steep cost. From 2001, when China joined the World Trade Organization, it increased its coal consumption by three times the original quantity—all to further production. While this move enabled China to fight rising poverty levels, the environmental consequences were dire.

Things reached the zenith in 2013 when Chinese citizens increasingly became aware of how difficult it had become to breathe normally on the streets. This led to a wake-up call, and China began recognizing that it could not continue consuming coal in this untoward manner. So, in 2013, the Government of China laid down a plan to invest over $270 billion to reduce air pollution by bringing down the use of coal and curbing vehicle emissions. This set the tone for China's participation in the negotiations of 2015 in Paris, where two hundred nations agreed to impose specific individual emissions targets subject to regular inspection, updates, and review. China pledged peak emissions by 2030, promising that after this, carbon dioxide emissions would decrease. The pledge came after the country had recognized that stark measures needed to be taken to improve the status of the climate as it was in the best interest of long-term national economics and geopolitical positioning.

From there, China boosted its reliance on green energy and clean fuel sources. Truth be told, the one-party system in China does leave room for authoritarian aggression at times. Still, it means that when a decision is taken, China can make the actions needed to solidify the decision with precision. China began decisive investment in green energy technology and infrastructure, including the billion-dollar subsidies to get its electric vehicle system up and running.

On the other hand, such a move in the United States would become immediately tied up in popular politics and partisan struggles, a large part of which often stems from the ridiculous belief that climate change is a myth. This belief, thankfully, has no place in China. In Chinese theory and practice, climate change is real, and it needs to be tackled. In China, the key is finding the balance between taking a risk and managing the finances needed to make that risk work. There have also been debates on how much responsibility and obligation should fall on China's shoulders since it is still a developing country. However, China has never contested the science behind climate change. That has been wholly accepted.

The Chinese administration also considers renewable energy a major economic investment area that can come with many profits. Fast forward to today, it is the world's leading producer, investor, and user of renewable energy sources. Many renewable energy variables —amounting to energy produced from about one out of three solar panels and one out of four wind turbines globally—come from China. It is also taking the lead in renewable energy sourced vehicles.

One has to remember that the size of China is vast, so while it may be making a lot of changes, it might still be a while before this gets translated to a visible impact. But that does not mean that it is not doing anything. Take, for instance, the use of renewable energy. About 13% of the energy being used in China comes from renewable sources, while the USA, even being a developed country, only sources 11% of its energy from renewable sources. The European Union as a whole generates 18% of its energy from renewable sources, and it includes a number of nations—which is woefully inadequate. From this vantage point, China is a much bigger contributor to environmental improvement.

The issue that China will need to look into is what it has been doing for the other countries. Half of the world's steel and cement is produced in China, and the government must consider what it can do to bring down the use of coal needed to fuel these sectors. These heavy industries rely heavily on heat-extensive emissions producing methods, which can be notoriously difficult and expensive to decarbonize. To do so, China will have to look for far more costly alternatives, such as green hydrogen. The cleanest fuel, natural gas, is also unavailable in China with abundance (as it is in the US).

Even with these restrictions, China made a pledge that it is working towards, while President Trump focused only on dissing China, not making any revelations on what the USA has done to benefit the environment.

The latter's speech also left no lasting impact. However, President Jinping's speech left a sizable impact worldwide, with countries like South Korea and Japan also following suit and pledging to reduce emissions. China's announcement served as the much-needed fuel for international negotiations surrounding the environment. International environmental politics aside, Jinping's declaration has also made a visible impact within China itself. Every ministry, company, state-dependent enterprise, and research organization must take the carbon neutrality pledge into due consideration and undertake measures to further it—because it is now a national goal. Already, the country is moving towards exploring developing sources of energy like green hydrogen and carbon capture and storage. Wind and solar power have already become major players in the race to reach peak emissions by 2030 and achieve carbon neutrality by 2060. What China is doing right now by investing in such sources is to build a market of competition surrounding them, which means that more and more people will now turn to these alternative sources because they know that the government is favoring them. This will take some reliance on the consumption of fossil fuels and redirect it to other, much environmentally safer alternatives.

While China continues to do all this, the USA is still trying to get its head back into the game. Four days after the Paris Agreement in 2016, Trump was elected as the president, and shortly after, he debunked the whole science of climate change, because according to him, it was cold outside when it was meant to be so. He also decided to move the country away from the Paris Agreement, because to him, it was unjust, and the developing countries should have played a bigger role in reducing the implications of environmental degradation. To quote his speech, "*Look at China, how filthy it is. Look at Rusia. Look at India. It's filthy. The air is filthy. The Paris Accord—I took us out because we were going to have to spend trillions of dollars, and we were treated very unfairly.*"

By the time the US withdrew from the accord, Trump had already abandoned about a hundred important laws that would have benefited the environment and weakened all rules that would have helped run fuel-efficient vehicles. While Biden is trying to undo this damage, it will not show any visible signs for a long while because of the sheer weight that the damage has taken on. It has cost a lot of critical time to get to this stage, and now to undo the deterioration of the previous administration, it will take a lot of financial and real-time effort. The rest of the world also did not abandon the pledge to improve the environment just because the USA left the accord. It has become a saving face kind of issue now—how will the USA undo all the damage it has left in the wake of leaving the accord? China did not leave the accord. China made commitments—and, as a matter of fact, every day, it moves closer and closer to meeting its 2030 goals.

The new American administration under Biden has much work to do if it wants to restore leadership in the arena of environmental politics. The first step will be to rejoin the Paris Agreement. The second will be to strengthen the existing commitments under the Paris Agreement. Initially, the US pledged to reduce emissions between 26 to 28% below its 2005 emissions levels by 2025. This goal, however, was seen as massively insufficient by Climate Tracker, an independent website that measures individual emissions by countries to keep overall global warming below 1.5 degrees Celsius.

Taken together, China and the US are two countries that contribute to about 40% of the World's GDP. They both must strive for goals to help achieve a sustainable environment while taking care of economic progress. China has already begun doing its part. When will the US catch up?

CHAPTER SUMMARY

- The availability of abundant labor, plentiful land, and lose laws that have enticed foreign investment have resulted in a significant portion of China's polluting output coming from other nations, such as the United States and Europe.
- China has invested about three trillion yuan in pollution control projects since 2012. The central government has shown its dedication to environmental preservation and social welfare via legislative reforms.
- To enhance water quality in key river basins and metropolitan areas throughout the state, in 2015, the state department also published a Plan Of Action for Prevention and Control of Water Pollution.
- China announced in June 2018 that it had established new objectives for collaborative management of greenhouse gas (GHG) emissions while simultaneously boosting air pollution control to attain a blue sky. China released the revised Action Plan in June 2018. MEE released an inclusive management plan for VOCs in key industries in 2012, which drew

attention to the problem of VOC emitted into the atmosphere.

- While 22% of all solar energy cells were made in the United States twenty years ago, just 1% are currently produced on American soil. The public had previously expected that major solar component manufacturers in the US would increase as the industry developed.
- Since 1965, the United States has produced the most greenhouse gas emissions of any country. The International Energy Agency estimates that the United States has emitted 278 billion megatons of carbon dioxide into the atmosphere during the past five decades, or approximately 21% of total global emissions during that period.
- A significant economic investment sector for the Chinese government is renewable energy, which can provide large revenues for the country. After all these years, it is now the world's largest source of renewable energy production, investment, and utilization. Approximately one-third of the world's solar panels and one-fourth of its wind turbines generate their electricity in China. It's also a pioneer in cars powered by clean, renewable energy.

- If the United States intends to regain its position as a world leader in environmental politics, the next government led by Joe Biden will have their work cut out for it. Rejoining the Paris Agreement will be the first step. The second step will be to reinforce the current pledges made as part of the Paris Climate Accord.

In the final chapter, you will learn what the United States can learn from China's example on the environment.

9

LEARNING FROM CHINA

From what we have studied this far, it is evident that China has moved ahead in terms of everything it has done to protect the environment. The USA has a lot of catching up to do. The first thing it has to do is honor the terms of the Paris Agreement. With the Copenhagen and Cancun accords, countries agreed to limit global temperature rises to 2 degrees Celsius over pre-industrial levels by the end of this century. The Paris Agreement restates the two-degree target while encouraging efforts to keep the temperature rise to 1.5 degrees Celsius or even less. The agreement also establishes two additional long-term mitigation priorities: first, a goal of reaching peak emission levels as soon as possible (while acknowledging that this will take more time for developing countries), and second, a goal of reaching net greenhouse gas neutrality by striking a balance between human activities that cause emissions and actions that prevent emissions by the second half of the century.

A review of country-specific pledges concluded that, although they get us close towards the two-degree target, they are not substantially ambitious to accomplish it. After conducting experiments, the Climate Action Tracker, a coalition of research institutes, determined that existing national mitigation and adaptation plans, if completely implemented, may still result in an increase in global temperature of 2.7 degrees Celsius by the turn of the era. Parties to the Paris Agreement are encouraged to create and communicate protracted reduced greenhouse gas emissions roadmaps. They were encouraged to submit such strategies to the United Nations Framework Convention on Climate Change by 2020. More than twenty nations have officially submitted their plans. According to the Alliance, the sixty-six countries that are members of the Climate Ambition Alliance have stated their intention to reach net-zero carbon emissions by 2050.

President Biden promised to reinstate the United States' participation in the Paris Agreement and give combating climate change a lot of emphasis in his administration. The key lies in restoring America's reputation and commitment to combat global warming. America's role is to ensure the world will increase its ambition to combat global warming due to America's steps to get to this stage. On February 19, 2021, the United States of America officially rejoined the accord. The findings of a recent study published in the journal Nature Sustainability demonstrate that as China's growing cities get richer, their greenhouse gas emissions per capita are all on a steady plane of decline. According to this research, this tendency may result in a general decrease in CO2 levels throughout the country, which would imply that, despite the existing goal for emissions to peak by 2030, China may, in reality, level off between 2021 and 2025.

To even consider keeping up, the United States should begin by upholding the promises that it has already made. Biden must return to Paris with a more aspirational commitment than ever, one that matches the increased ambitions established by the European Union and the United Kingdom and meets the needs and expectations of domestic environmental groups. Furthermore, if the United States makes big promises in its Nationally Determined Contributions, it should be a powerful encouragement for all other nations to follow suit.

Most significantly, though, a more ambitious carbon reduction goal for the United States must be attainable. Professor of energies and ecological policy at the Fletcher School at Tufts University, Kelly Sims Gallagher, has noted that the current administration would need some time to find out what policy and legislative options are available to address climate change (Gallagher, 2012). Taking every step possible to guarantee that the United States can achieve the goal will be essential.

However, enacting climate legislation soon would help convince allies that the United States is serious about returning to the climate battle. It will perhaps inspire other nations to take decisive actions; however, the United States should be modest in its first steps. An essential first step is a concentrated focus on domestic affairs and obtaining the consent of the populace, which would then provide the country a more vital spot in terms of the international policy on climate change. To learn from China, it should move towards reaching a net-zero emissions target—but only *after* it has figured out a good policy on how to go about this, something that will resonate not just with the administration but also the people of the US at large. For a federal government like that of the US, this is of utmost importance.

The United States must commit to long-term, large-scale public investments in education, affordable healthcare, energy, and construction to learn from China. Because renewable energy is beneficial to both the economy and the environment, it should take time to consider implementing such an investment in the best way possible. In addition, the United States should consider fostering the export of renewable energy knowledge to other nations. China has spent tens of billions of dollars on renewable energy technologies in Australia, Germany, Brazil, Egypt, and other countries.

We looked at how China is pushing the deployment of electric cars across the nation and found some interesting results. To catch up and fulfill its goals, the United States must also consider adopting vehicles that operate on renewable energy resources, which it may accomplish via tax breaks and substantial subsidies.

Each time the United States has faced an environmental crisis, the nation's ability to protect the environment has increased. While developing a solid history of locally driven conservation advocacy, the United States also educated many academics and ecological specialists and worked to create and strengthen its environmental laws. Even more significantly, the beauty and biodiversity that the country has preserved have now become a cultural touchstone. America's desire to protect land, water, and wildlife for future generations, despite the severe partisanship of today's political climate, unites people of all political parties and beliefs.

In the right hands, these shared values—when harnessed through accessible legal, outreach, systematic, and electoral channels—can empower current generations. The current generation can successfully restrict biodiversity extinctions, combat anthropogenic global warming, minimize pollution problems, and shield healthy natural systems on which subsequent generations will rely. A clear and shared perspective is absent from the country's political discourse on how much ecology the nation wants to preserve, in what shape, at what expense. Consequently, the United States is substantially underplaying its ability to safeguard the national environment. This is where it has to learn from China—where protecting the environment has become the watchword for the government. The Chinese administration looks at its national environment as a priceless resource that must be improved and safeguarded. If the USA is to catch up, it has to adopt the same strategy.

The United States must take its responsibilities in maintaining a clean environment seriously. Policies that help lower the use of coal and oil without carbon capture and sequestration should be highlighted, as should ensuring that petroleum and gas manufactured in the United States emits as little carbon as possible. It should also prepare for the inevitable decline of national oil and gas production by enacting policies that will assist affected regions through the transition. While it is okay to oppose less aggressive climate policy, one also must be extremely cautious of supporting policies that concentrate their energies on the consumption and emissions sides of production—particularly since, in some cases, oil and gas continue to be the easiest sources available. Much research needs to be done before the administration can take the call to stop consuming and emitting oil and gas altogether. Falling prices for sustainable power assist in displacing fossil fuels in electricity production, but oil is tougher to replicate in other applications.

To promote the shift in the most challenging areas, research and technology will be required. The USA needs a long-term plan and vision to allow this to happen—it must work together and build a shared vision based on "we" and not "I." The US administration must look at climate change as a problem affecting the whole world and mitigate it in partnership with other nations. US leaders have to believe that by working together, while achieving the net result is complex and will take time—it will undoubtedly get there. China began as one of the most polluting countries, and look at where it is now. The USA must believe that it can get there as well.

After China established itself as a global powerhouse in renewable energy, the Biden presidency has focused on countering that dominance by increasing the competitiveness of American businesses. China is quickly expanding, and while coal and fossil fuels account for the majority of its growth, the Chinese government has been using a combination of financial incentives and government regulations to encourage the development of the clean energy industry. Although renewable energy financing in the United States is insufficient to compete with fossil fuels on its own, it has the potential to assist strategically important nations in developing clean energy markets. Renewable energy prices are falling fast as design engineers compete in the market and gain experience through trial and error. Rising energy prices enable developing nations to fulfill their energy requirements reasonably while simultaneously increasing energy security. That industry is not only expanding inside China, but it is also expanding internationally. It is having a significant impact on the markets in the United States and Europe.

The scenario in China is diametrically opposed to the approach to renewable energy that the current government has adopted in the United States of America.

Many of the United States' attempts to encourage clean technology are handled unexpectedly, with subsidies and rewards that lapse and then reincarnate and then expire again, and an absence of clarity in terms of a guiding policy that would enable businesses to prepare for the future of the sector. This inconsistency has placed the United States at a competitive disadvantage when vying against China for a share of renewable energy markets, both domestically and internationally. The United States should let the competitive market determine the fate of local companies to make amends and catch up. However, it must also incorporate long-term industrial plans and the environmental requirements of competing nations, in addition to the hidden costs of the current energy system. This will ensure the presence of long-term commitments that local companies have to adhere to, making it easier for the administration to know if they are engaging in harmful environmental practices.

When nations with strong industrial policies, such as China, are placed against inconsistent and unpredictable policies like those in the USA, the latter will have a hard time competing on an equal playing field. Many American businesses and sectors must depend on trade policy to preserve their competitive interests since there are no clear and compelling policies to promote the development of environmentally sustainable technology in the United States. Trade enforcement is essential, but it is even more critical to have a solid foundation of domestic policy on which to build upon in the first place. The United States and China have a shared interest in the deployment of renewable energy. While both countries may be in a position of competing with one another for new markets, a portion of their attention should be directed on determining what the two nations can do collaboratively to help accelerate the implementation of these technologies.

The Chinese project members and funders will have significant incentives to fulfill the demand to reduce China's industrial excess as long as nations with increasing energy demands continue to invest in coal, natural gas, and oil, among other energy sources. Rather than just handing the ball over to China as a lender and shareholder in foreign oil and coal initiatives, the United States should use a two-pronged approach to encourage China to engage more heavily in clean energy financing and infrastructure. Emerging markets are very vulnerable to the price of energy, and the International Development Fund Corporation (DFC) of the United States may provide cheap loans to finance sustainable energy projects in these nations.

Such loans have the potential to replace fossil fuels while also contributing to broader climate change mitigation efforts. The Development Finance Corporation (DFC) has extensive expertise in infrastructure financing in developing markets and may loan up to fifty-sixty billion dollars under current regulations. Clean energy initiatives that decrease greenhouse gas emissions and assist developing nations in avoiding carbon-intensive growth pathways should be supported by the DFC, which the United States should establish.

The United States may be able to influence markets in a more environmentally friendly direction due to this. As renewable energy costs continue to decline, governments will no longer have significant financial incentives to emphasize fossil fuels over other energy sources. Supporting renewable energy will decrease host-country reliance on fossil fuels, even though the United States cannot take the toll of financing the whole energy needs of developing nations. This means that clean energy financing from the United States may also help green Chinese overseas credit. As developing countries grow their clean energy markets, they will look for loans to fund these projects, resulting in new investment possibilities for investors.

Ironically, the United States is doing all it can to squander its own economic, technical, and geopolitical advantages at the same time that China rises economically and geopolitically. With the avarice of its financial elite, the United States has become a target of its own political system. Their narrow goals are to reduce business and individual tax rates and severely restrict positive and productive involvement in global economic development. They are so dismissive of US humanitarian assistance that they may have opened the door to China's emergence as the world's leading provider of development financing. To learn from China, the United States can look at initiatives like assisting developing nations in controlling energy projects by providing financial assistance. Only a few developing countries have the organizational capacity to track and enforce sustainability measures for energy projects.

For instance, in Indonesia, ecological controllers at both the national and local levels suffer from a severe understaffing issue, harming their ability to conduct environmental monitoring. Consequently, many dirty-energy (coal and oil) driven power projects can begin operations even before the necessary environmental impact studies are finished. As a result, when developing nations lack a sufficiently strict environmental capability, losses such as carbon dioxide emissions cannot be adequately addressed. Suppose the United States can assist these nations in increasing their capacity. In that case, clean energy may gain an edge over fossil fuels by accounting for the financial costs associated with negative environmental externalities, as opposed to fossil fuels.

Fortunately, it appears that cooperation is on the horizon. At the invitation of Joe Biden, Chinese President Xi Jinping attended a United States-led climate change summit on Thursday, April 20, 2021, marking the first session between the two leaders since the start of the new American administration in January 2017. Following the United States' re-entry into the 2015 Paris Agreement on global carbon emissions reductions, Biden invited numerous world leaders to participate in the two-day virtual summit in early November of 2021. President Xi issued a brief statement to the climate conference at its opening instead of attending in person or pre-recording a video address. His written statement called for an acceleration in actions on global climate change.

Beijing and Washington have been at odds over various issues for years, ranging from accusations of human rights violations to China's economic strength over other countries, among others. The first increased in-person talks between the United States and China occurred in Alaska in March 2021; they were defined by acrimony and the failure to achieve any diplomatic breakthroughs. However, the two countries seem to have rediscovered a shared interest in combating climate change. This may have something to do with the roles that each of them has played in the emission of copious amounts of greenhouse gases into the environment. It appears that the USA is now willing to learn from China to stave off the damage it inflicted for all these years.

International energy cooperation between the United States and China is essential for global energy security, climate change mitigation, and nuclear security concerns. It is a top priority for both countries. The United States and China are working together on cementitious, group efforts that advertise reliable and sustainable power, resource efficiency, energy security, reducing greenhouse gas emissions, and reducing electricity costs.

The Department of Energy's Office of Global Affairs oversees a strong relationship with China on renewable energy cooperation, which has reached significant milestones in recent years. Examples include collaboration on various renewable energy projects, such as the United States-China Clean Energy Research Center (CERC) and the Energy Efficiency Action Plan (EEAP). DOE is concentrating on leveraging the strong and extensive bilateral cooperation on clean energy between the United States and China to lead the way in expanding markets and developing solutions to energy problems. These solutions result from collaboration on a properly integrated optimization and innovation chain, with the goals of reducing energy consumption, supporting game-changing breakthroughs, driving down the cost of energy-efficient technologies, and deploying those innovations at scale. The Department of Energy collaborates with China to use sustainable energy technologies to enhance development, security, and prosperity. This kind of cooperation may have been impossible even a few years back—but given the urgency of climatic situations today, it has become an imperative.

The United States and China consider the safe and secure production of nuclear power to be a contingency. This will significantly reduce emissions. Both countries must give due consideration to nuclear energy's role in achieving global carbon reductions. They are also collaborating to guarantee that China joins the United States and other major nations as a leader in nuclear security. For example, following years of strategy and development cooperation, the United States and China established the Institute for Nuclear Security in Beijing in 2016. The Center will sponsor educational programs for Chinese authorities, regional partners, and other partner countries to raise global awareness and engagement in nuclear security.

Energy cooperation between the United States and China is carried out via various bilateral forums, including the CERC, EEAP, Clean Energy Working Group, Oil and Gas Industry Forum, and Clean Coal Industry Forum. As part of the energy cooperation with China, the USA participates in international discussion forums such as the International Energy Agency, the Clean Energy Ministerial (CEM), Mission Innovation, and others. Wind and solar power are only the beginning of the clean energy technologies that exist today. Carbon capture, utilization, storage, and nuclear power are also new avenues of cooperation that can help protect the universal environment while helping both nations move towards net-zero emissions.

So, in the end, the USA must begin its journey of retribution by accepting the current condition that its environment is in. Without that, every step it takes will be in vain. It has to keep its ego aside and look to avenues of cooperation and collaboration with China to benefit the world by cleaning up the environment and massively benefit in terms of ecological politics and financing. Aside from that, the Chinese administration will run the danger of destroying its image as a competent major power and will be under increasing pressure to impose tighter restrictions on Chinese companies' international ventures. Building capacity in developing markets will push China to examine its financial institutions and project developers thoroughly. So, if the USA moves in these directions, it will be good for the country and the whole world. At the end of the day, the world has to look at the environment's problems as a collective whole. It cannot keep on polluting with no accord for accepting responsibilities—because when people suffer the consequences of a bad environment, they suffer as a whole species—not just as an American or Chinese resident. The concern is global—and perhaps, the road to a cleaner future is in the hands of these two powerhouses to collaborate and find a long-term solution.

CHAPTER SUMMARY

- According to what we've learned thus far, China has made significant strides in environmental protection. The United States must make up ground quickly. As a first step, it must uphold its end of the bargain under the Paris Agreement. It reiterates the two-degree goal while urging efforts to limit global warming to 1.5 degrees Celsius or below.
- Both are achieving peak emission levels as soon as feasible, and achieving net greenhouse gas neutrality by the end of the century are goals outlined in the agreement. Human activities

that contribute to emissions must be coordinated with actions that lead to their removal.

- Paris Agreement Parties are urged to develop and publish long-term plans for cutting greenhouse gas emissions, which they should do by 2020 to be included in a UN climate change convention.
- The United States of America must take its environmental duties seriously. To reduce coal and oil usage without using carbon capture and sequestration, it's important to emphasize policies that reduce petroleum and gas production emissions in the United States.
- The most critical step is a laser-like concentration on local issues and accomplishments, putting the country better positioned to influence international climate change policy. To learn from China, the United States should certainly work toward a net-zero goal, but only after devising a strategy that resonates with both the government and the American people as a whole.
- This country has to think about assisting other countries in acquiring expertise in renewable energy sources. Renewable energy technology in nations like Australia, Germany, Brazil,

Egypt, and others has fetched China tens of billions of dollars in investment from the Chinese government.

- We investigated China's efforts to promote electric vehicles throughout the country and came up with some surprising findings. There are tax incentives and significant subsidies available to encourage Americans to buy cars that run on renewable energy sources, which will help the country catch up and meet its objectives.
- The deployment of renewable energy is something that both the United States and China are concerned about. There may be some competition for new markets, but both countries should focus on what they can do together to speed up the adoption of these technologies, not only in terms of market share.
- The first round of enhanced in-person discussions between the United States and China took place in Alaska in March 2021, and they were marked by antagonism and the refusal to form any diplomatic breakthroughs, according to the United States. On the other hand, the two nations seem to have regained a common interest in fighting global climate change.

- Held by The United States' re-entry into the 2015 Paris Agreement on carbon pollution emissions reductions, Vice President Joe Biden asked many world leaders to participate in a two-day virtual summit, which took place over two days in early November. President Xi issued a brief statement to the climate conference instead of attending in person. His written statement called for an acceleration of global climate action.
- It is possible that clean energy funding from the United States will assist green Chinese overseas credit to grow. With the development of emerging nations' clean energy markets, they will begin to seek loans to help them finance these projects, leading to new investment opportunities for financial institutions.
- The United States could create clean energy initiatives that reduce greenhouse gas emissions while also assisting emerging countries in avoiding carbon-intensive development paths.

FINAL WORDS

With this, we come to the crucial question: why is protecting the environment so important? First and foremost, we should all be familiar with the word "ecology." All living creatures live within their own advantageous sectors of the ecology. As a result, the ecosystem cannot be simply ignored if proper care is not taken to maintain it. Indeed, the Lord has established a degree of adaptability in the ecosystem for all creatures that live on the planet. Human involvement has caused significant damage to many ecosystems, although the environment may persist in its natural form for a long time. It is critical that we preserve our environment to ensure that we can continue to live on a safe, healthy planet.

Nine out of ten people on the planet are inhaling polluted air, which is detrimental to their health and shortens their life expectancy. The number of people who die from illnesses and infections linked to air pollution each year is about 7 million, more than five times the number of people who die in road traffic accidents.

In addition, exposure to pollutants may affect the brain, resulting in developmental delays, behavioral issues, and even lower IQ scores in children. Among the elderly, pollutants have been linked to the development of Alzheimer's and Parkinson's illnesses.

People's increased interaction with wildlife due to the modification of land to make way for houses, farms, and industries has increased the chances of diseases spreading from feral animals to humans. According to estimates, animal-borne diseases account for 60% of all human infections. There is also a slew of additional viruses ready to make the leap from animals to people.

In addition to living animals, we should be concerned about protecting the environment's plants and trees, which offer us even more significant advantages directly—they supply us with the oxygen we need to survive in this world. Countries in industrialized and developed economies are likewise at risk of being contaminated and adversely impacted by the negative environmental effect caused by human activity.

Human diets have grown over 30% more similar in the past five decades alone. Today, almost one-third of the world's population suffers from some kind of malnutrition. A large proportion of the world's population is afflicted by diet-related illnesses such as heart disease, diabetes, and cancer.

Waste dumped in our water sources is wreaking havoc on aquatic life, and pollution from industries, our use of plastic bags, and cars contribute to climate change, amongst other things. Many of the Earth's resources are being drained by humans. With the consequences posing a threat to the lives of millions of people across the world. To mitigate the negative effect of human activities on the environment and the climate, many international accords have been negotiated and ratified by governments worldwide. However, the goals established during these agreements are often not reached, and the consequences of climate change are left unabated. It is beneficial to some degree.

Governments must safeguard ecosystems that are critical in the battle against climate change. Rivers, marshes, seas, forests, and mangroves all absorb significant amounts of carbon, which helps to decrease global warming and reduce greenhouse gas emissions.

Mangroves also create a barrier against severe cyclones, and wetlands absorb surplus water from flooding, both of which are extreme weather phenomena becoming more frequent due to global warming. In contrast to large factory farms that use unsustainable practices and do not care about land restoration, small local producers benefit from government support and help make crops and animals more resilient to climate change. Unlike huge factory farms, small local producers utilize sustainable practices and care about revegetation, benefiting nearby communities and helping to make animals and crops more adaptable to climate change. It's less about everyone becoming vegetarians and more about supporting people who grow our food in an environmentally friendly way, respectful to the environment.

All of us will gain if we each make our own little pledges to safeguard nature and to fight against negative environmental effects. This would enable us to reestablish a positive connection with nature in our lives. A cleaner environment will reduce human health issues such as respiratory illnesses, heart problems, illnesses, and death caused by pollution in our environment, which would also benefit the ecosystem. Both China and the United States have a long way to go before they can be considered successful. One can only hope that the two countries will enter into more environmental agreements in the future and that America will learn from the example that China has already set. I hope that the two countries will work together to uplift and protect the environment while also caring for their respective national advancements.

Dear readers, I hope you found this read to be a turning point for you. The changes in the environment are a reality, and the sooner we acknowledge it, the sooner we can hope to protect and reverse the damage that has been done. If this read gave you perspective, don't forget to leave a review on Amazon.

REFERENCES

Bombardini, Matilde, and Bingjing Li. 2020. Trade, Pollution and Mortality in China. Journal of International Economics,103321.

Electricity consumption in China between 2010 and 2020(in terawatt hours). (2021). Statista. https://www.statista.com/statistics/302203/china-electricity-consumption/

Gallagher, K. (2012). Breaking the Climate Impasse with China: A Global Solution. *Tufts University*.

Gallagher, K. S., & Zhang, F. (2019). China is positioned to lead on climate change as the US rolls back its policies. The Conversation. https://theconversation.com/china-is-positioned-to-lead-on-climate-change-as-the-us-rolls-back-its-policies-114897

Gardner, D. K. (2014). China's Off-the-Chart Air Pollution Why It Matters (and Not Only to the Chinese) - Part One. The National Bureau of Asian Research.

Greenstone, M., He, G., Li, S., & Zou, E. (2021). China's War On Pollution: Evidence From The First Five Years (No. 28467).

"Global Burden of Disease Study 2010," Lancet, December 13, 2012, http://www.thelancet.com/themed/global-burden-of-disease.

He, Guojun, Maoyong Fan, Maigeng Zhou. 2016. The Effect of Air Pollution on Mortality in China: Evidence from the 2008 Beijing Olympic Games. Journal of Environmental Economics and Management 79: 18-39.

Kahn, J., & Yardley, J. (2007). As China Roars, Pollution Reaches Deadly Extremes. New York Times. https://www.nytimes.com/2007/08/26/world/asia/ 26china.html

Karplus, V. J., Zhang, S., & Almond, D. (2018). Quantifying coal power plant responses to tighter SO2 emissions standards in China. Proceedings of the National Academy of Sciences of the United States of America, 115(27), 7004–7009. https://doi.org/https://doi.org/10.1073/ pnas.1800605115

King, H. (2019). What Is PM 2.5 and How Can You Reduce Your Exposure? Molekule. https://molekule.-science/what-is-pm-2-5-and-how-can-you-reduce-your-exposure/

Mac, P. (2015). The History of Air Pollution in China. Colgate University. https://chinaenv.colgate.edu/air-pollution/air-pollution-in-china/

Maizland, L. (2021). China's Fight Against Climate Change and Environmental Degradation. Council on Foreign Relations. https://www.cfr.org/back-grounder/china-climate-change-policies-environmen-tal-degradation

Myllyvirta, L. (2020). Influential academics reveal how China can achieve its carbon neutrality goal. Carbon Brief. https://www.carbonbrief.org/influential-acade-mics-reveal-how-china-can-achieve-its-carbon-neutrality-goal

Nightingale, A. (2019). Forget Tesla, It's China's E-Buses That Are Denting Oil Demand. Bloomberg Quint. https://www.bloombergquint.com/china/forget-tesla-it-s-china-s-e-buses-that-are-denting-oil-demand

Stoerk, Thomas (2015). Statistical corruption in Beijing's air quality data has likely ended in 2012.

Atmospheric Environment, (), S1352231015306336–. doi:10.1016/j.atmosenv.2015.12.055

Tanaka, Shinsuke. 2015. Environmental Regulations on Air Pollution in China and Their Impact on Infant Mortality. Journal of Health Economics 42: 90-103.

Xu, M., Stanway, D., & Navaratnam, S. (2020). China's new green development fund raises $12 billion in phase 1. Reuters World News. https://www.reuters.com/article/us-china-environment-idUSKCN24T0E0

Yuyu Chen, Avraham Ebenstein, Michael Greenstone, and Hongbin Li, "Evidence on the Impact of Sustained Exposure to Air Pollution on Life Expectancy from China's Huai River Policy," Proceedings of the National Academy of Sciences of the United States of America 110, no. 32 (2013): 12936–41, http://www.pnas.org/content/110/32/12936.abstract.

Made in the USA
Monee, IL
15 January 2022

89003069R10111